Industrial Design, Competition and Globalization

Also by John R. Bryson

THE HANDBOOK OF SERVICE INDUSTRIES (*with P. W. Daniels*)

SERVICE INDUSTRIES IN THE GLOBAL ECONOMY (*with P. W. Daniels*)

SERVICE WORLDS: People, Organisations, Technologies (*with P. W. Daniels and B. Warf*)

The Front Cover

The Acme Thunderer. Designed and made in Birmingham, England, by J. Hudson and Co. (Whistles) Ltd., Birmingham.

The Morgan Aero SuperSports. Drawn by Matthew Humphries and designed and engineered for production by the "Morgan Design" team, Morgan Motor Company, Malvern Link, England.

The Tripp Trapp Chair. Designed by Peter Opsvik and manufactured by Stokke Fabrikker AS. This product was awarded the Norwegian Design Council Classical Award in 1995.

The ULSTEIN X-BOW®. Designed by the Ulstein Group. Awarded the Annual Feat of Engineering Award in 2005, the Ship of the Year Award by Skipsrevyen in 2006, and also nominated for a Sea Trade Award. The ship is owned by Bourbon Offshore, Norway.

Industrial Design, Competition and Globalization

Edited By

Grete Rusten
Professor, Department of Geography, University of Bergen

and

John R. Bryson
Professor of Enterprise and Economic Geography, University of Birmingham

palgrave
macmillan

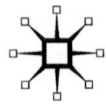

Selection and editorial matter © Grete Rusten and John R. Bryson 2010
Individual chapters © the contributors 2010
Foreword © Jan R. Stavik 2010

All rights reserved. No reproduction, copy or transmission of this publication may be made without written permission.

No portion of this publication may be reproduced, copied or transmitted save with written permission or in accordance with the provisions of the Copyright, Designs and Patents Act 1988, or under the terms of any licence permitting limited copying issued by the Copyright Licensing Agency, Saffron House, 6-10 Kirby Street, London EC1N 8TS.

Any person who does any unauthorized act in relation to this publication may be liable to criminal prosecution and civil claims for damages.

The authors have asserted their rights to be identified as the authors of this work in accordance with the Copyright, Designs and Patents Act 1988.

First published 2010 by
PALGRAVE MACMILLAN

Palgrave Macmillan in the UK is an imprint of Macmillan Publishers Limited, registered in England, company number 785998, of Houndmills, Basingstoke, Hampshire RG21 6XS.

Palgrave Macmillan in the US is a division of St Martin's Press LLC, 175 Fifth Avenue, New York, NY 10010.

Palgrave Macmillan is the global academic imprint of the above companies and has companies and representatives throughout the world.

Palgrave® and Macmillan® are registered trademarks in the United States, the United Kingdom, Europe and other countries.

ISBN: 978–0–230–20349–5 hardback

This book is printed on paper suitable for recycling and made from fully managed and sustained forest sources. Logging, pulping and manufacturing processes are expected to conform to the environmental regulations of the country of origin.

A catalogue record for this book is available from the British Library.

A catalog record for this book is available from the Library of Congress.

10 9 8 7 6 5 4 3 2 1
19 18 17 16 15 14 13 12 11 10

Printed and bound in Great Britain by
CPI Antony Rowe, Chippenham and Eastbourne

Contents

List of Illustrations	vii
List of Contributors	ix
Acknowledgements	xi
Foreword	xiii

1. Industrial Design, Competitiveness, Globalization and Organizational Strategy   1
   Grete Rusten and John R. Bryson

Part I Industrial Design and National Competitiveness

2. Industrial Design, National Competitiveness and the Emergence of Design-Centred Economic Policy   23
   John R. Bryson

3. Better by Design? A Critical Appraisal of the Creative Economy in Finland   50
   Nikodemus Solitander

4. Locational Patterns and Competitive Characteristics of Industrial Design Firms in the United States   81
   Alan D. MacPherson and Vida Vanchan

5. The Geography of Producing and Marketing Design for Montreal Fashion: Exploring the Role of Cultural Intermediaries   93
   Norma Rantisi

6. Designed Here, Made There? Project-based Design Work in Toronto, Canada   117
   Tara Vinodrai

7. The Geography of Design in the City   141
   Thomas A. Hutton

Part II Design and Firm Competitiveness

8. Design and Gender with a Competitive Edge   169
   Lena Hansson, Magnus Mörck and Magdalena Petersson McIntyre

9	Riding the Waves of Design: Industrial Design and Competitive Products in the Norwegian Marine, Maritime and Offshore Sectors *Grete Rusten*	195
10	Is Good Design Good Business? *Gorm Gabrielsen, Kjell Grønhaug, Lynn Kahle, Tore Kristensen, Thomas Plenborg and Ricky Wilke*	220

Index   243

Illustrations

Tables

2.1	Selected International Exhibitions held during the reign of Queen Victoria (1837–1901)	44
4.1	Top ten states in terms of employment counts (2005)	83
4.2	Top ten metropolitan areas (establishment counts) and metropolitan population sizes (2005)	84
4.3	Top ten metropolitan areas: Establishments and market specializations (2005)	85
5.1	Largest ten retailers' share of total Canadian apparel market, retail dollar sales (January–December 2006)	99
6.1	Size distribution of establishments in Canada's specialized design services industry (1998–2005)	125
6.2	Design employment and location quotients for major Canadian cities (2001)	126
6.3	Evidence of outsourcing relationships in Canada	129
10.1	Overview of design-performance studies	225
10.2	Rank correlation matrix, Spearman's rho	235
10.A.1	Financial data	238
10.A.2	Rating scales for the three qualities	239

Figures

3.1	Creative industries: Empowering the centre	61
5.1	Overview of the structure of the Montreal fashion industry: from production to consumption	100
5.2	Canadian apparel market – domestic and imports (1995–2005)	100
5.3	Canadian apparel market shipments (1995–2005)	101
6.1	Employment in the Canadian design sector: Industry vs. occupation (1987–2004)	124
6.2	The growth of Canadian design activity (1987–2004)	124
6.3	Expenditure on contract design work in Canadian firms (1998–2002)	130
7.1	Schematic of applied design clusters within the metropolitan space economy	146

7.2	Major service clusters in the polycentric global city	151
7.3	London's 'City Fringe' districts	152
7.4	Specialized industrial zones and sites, Clerkenwell, London Borough of Islington	155
7.5	Design industries and firms, Bermondsey Street (Southwark)	157
7.6	Specialized production areas in Vancouver's central area	160
8.1	Workwear Kilt	182
8.2	Power Drill	185
8.3	Volvo YCC	187
9.1	Distribution of designers by sector (Statistics Norway – unpublished data)	209
9.2	The Autochief ® C20 propulsion control system for maritime vessels	211
9.3	The ULSTEIN-X-BOW® Containership	212
9.4	LNG fuelled passenger ferry	213
10.1	Logos from Royal Greenland and COOP compared with respect to 'Functionality', 'Credibility' and 'Expressivity'	232
10.2	Products from Nomeco and Novo Nordisk compared with respect to 'Functionality', 'Credibility' and 'Expressivity'	233

List of Contributors

Editors

Rusten, Grete Professor, Department of Geography, University of Bergen, and Honorary Research Fellow, School of Geography, Earth and Environmental Sciences, The University of Birmingham UK.

Bryson, John R. Professor of Enterprise and Economic Geography, Head of Society, Economy and Environment Research Group, School of Geography, Earth and Environmental Sciences, The University of Birmingham, Edgbaston, Birmingham, UK.

Other contributors

Gabrielsen, Gorm Associate Professor, The Statistics Research Group, Copenhagen Business School.

Grønhaug, Kjell Professor, Department of Strategy and Management, The Norwegian School of Economics and Business Administration, Bergen.

Hansson, Lena PhD, School of Business, Economics and Law, and researcher at the Center for Consumer Science (CFK), University of Gothenburg, Sweden.

Hutton, Thomas A. Professor, Centre for Human Settlements, and Associate Director, School of Community & Regional Planning, University of British Columbia, Vancouver.

Kahle, Lynn PhD student, The Marketing Department, Copenhagen Business School.

Kristensen, Tore Associate Professor, The Marketing Department, Copenhagen Business School.

MacPherson, Alan D. Professor, Department of Geography, and Director, Canada-United States Trade Center, University at Buffalo, US.

Mörck, Magnus Associate Professor, Centre for Consumer Science, School of Business, Economic & Law, University of Gothenburg, Sweden.

Petersson, Magdalena McIntyre PhD, Centre for Consumer Science, School of Business, Economic & Law, University of Gothenburg, Sweden.

Plenborg, Thomas Professor, The Accounting Department, Copenhagen Business School.

Rantisi, Norma M Associate Professor, Department of Geography, Planning and Environment, Concordia University, Canada.

Solitander, Nikodemus Ph.D. Candidate, Supply Chain Management and Corporate Geography, Department of Marketing, Hanken School of Economics, Helsinki, Finland.

Vanchan, Vida Assistant Professor, Department of Geography and Planning, School of Natural and Social Sciences, Buffalo State College, US.

Vinodrai, Tara Assistant Professor, Department of Geography and Environmental Management and Centre for Environment and Business, Faculty of Environment, University of Waterloo.

Wilke, Ricky Associate Professor, The Marketing Department, Copenhagen Business School.

Acknowledgements

Producing an edited book may appear at first sight to be a comparatively simple task. All one has to do is identify a title, a set of aims and objectives, a structure and a possible list of chapter titles or topics. The next task involves persuading academics to participate in the project and finally the identification of a publisher willing to produce the book. Writing is followed by editing and then more writing and finally the delivery of the manuscript to the publisher. If only the production of an edited book were as simple as this description. This book began life at a meeting in Bergen, Norway, where we discussed and agreed on proposals for two linked books on design. Our initial interest in this topic developed from a research project on Design Competitiveness funded by the Norwegian Research Council in 2006–07. One proposal was for an authored book and one was the proposal that began the journey that eventually led to the publication of this edited collection. During the course of the day we had an interesting and stimulating intellectual discussion and hoped that one day that our deliberations would result in the publication of an edited book that would begin to address the relationship between design and competitiveness.

Producing this edited book involved numerous meetings of the two editors, constant e-mailing and the cooperation of the chapter authors and the publisher. We would like to thank Virginia Thorp, the Senior Commissioning Editor, Business and Management at Palgrave Macmillan for commissioning the book and Paul Milner, Editorial Assistant at Palgrave Macmillan for the care and attention he has given to managing the book's travels through the publishing production process.

An edited book is a partnership between an editor, in this case two editors, and a group of chapter authors. As editors we would like to thank the chapter authors for rising to the challenge that we set them and for working with us in a constructive manner over the last few months. We would also like to thank Anne Kristin Wilhelmsen at the Institute for Research in Economics and Business Administration (SNF), Bergen, who played an extremely important role in integrating the various computer files together and in ensuring that we delivered a relatively clean and integrated manuscript to the publisher.

Developing a proposal for an edited book represents the start of a long and interesting journey, and unexpected things can happen along the way. It is therefore with great regret that we have to report the passing of two of our chapter authors and colleagues who made important contributions to this book: Professor Magnus Mörck, Centre for Consumer Science, School of Business, Economic & Law, University of Gothenburg, Sweden and Professor Alan D, MacPherson, Department of Geography, and Director, Canada-United States Trade Centre, University at Buffalo, US. May you rest in peace. Our thoughts are with you and your families.

John R. Bryson, Birmingham, UK
Grete Rusten, Bergen, Norway
14 May 2009

Foreword

Even though we are in the middle of a very serious global economic downturn, competitiveness will continue to be a major challenge for all international companies, their management and their boards. It is therefore of vital importance that managers and board members are aware of the innovative potential which design represents. As the Norwegian Minister for Trade and Industry, Sylvia Brustad, said at the opening of NDC's Design for Business Conference 2008: "Companies must increase their awareness – and use of design – for increased innovation and competitiveness. This must become a focus area for all company boards."

In line with the above, the Norwegian Design Council keeps inspiring and assisting companies in their efforts to become more proficient in using design as a strategic tool for innovation, in order to obtain increased competitiveness. In this context it is very important that Norway's first National White Paper on Innovation, which was launched by MITI in December 2008, contains a separate chapter on design as a tool for innovation, and also introduces a National Program on Design-driven Innovation.

This particular book approaches the task of recognizing the role that design plays in our globalized economy, and it may encourage scholars, practitioners and policy makers to further focus on this issue.

Striving for competitiveness will always be a challenge; design is one way of getting there.

Jan R. Stavik, Managing Director,
The Norwegian Design Council

1
Industrial Design, Competitiveness, Globalization and Organizational Strategy

Grete Rusten and John R. Bryson

Introduction

The ongoing deepening of globalization is associated with enhanced competition and especially price-based and design-based competitiveness. High value-added goods are developed, designed and still manufactured in developed market economies whilst low value-added goods may be designed in developed market economies, but manufactured in low-cost economies like China, India or Turkey (Deitz and Orr 2006). The concentration of the production of low value-added manufactured products in countries like China represents a stage in the evolution of the economic geographies of production. Nevertheless, Chinese firms are trying to move up the value chain to produce design-intensive and branded high value-added goods. The relationship between high value-added goods and developed market economies is partly explained by a shift that has occurred in these locations, away from competition based on price, to competition based on intangibles such as design, branding, research and development (R&D) and embedded or attached services (Bryson et al. 2004; Bryson 2008, 2009a & b). It is important not to be too fixated on consumer products. The majority of products are designed; this includes machine tools, ships, medical scanners and safety clothing intended for extreme environments and used, for instance in marine fishfarming or the oil and gas industry (Rusten and Bryson 2007, Rusten and Stensheim 2007, Rusten 2008).

The emphasis that is being placed on investment in intangibles represents an awareness by some companies located in high-cost locations that competing solely on price is increasingly impossible (Bryson et al. 2008). The alternative may involve shifting manufacturing production and even some service functions to low-cost locations (Bryson 2007) and leaving a slimmed-down management, marketing and R&D facility in a developed market economy. Investment in intangibles now matters and is considered by governments to be a critical element of national competitiveness. Nevertheless, investment in intangibles is extremely difficult to measure and also difficult to address through policy development. In 2008, the British government published a new manufacturing strategy that identified five interrelated dynamics that continue to reshape global manufacturing:

- the increasing complexity of global value chains,
- the accelerated pace of technology exploitation,
- the growing importance of investment in intangibles,
- the increased recognition that investment in people and skills is amongst the most important investments for firms to make, and
- the move to a low-carbon economy (BERR 2008: 12).

The emphasis placed on intangibles recognizes the importance of knowledge assets that include 'design and other aspects of product development; software; brand-building; training; and improvements to business processes. Such investment boosts firms' competitiveness and enables products to meet changing consumer needs. The government's role is to ensure companies have the right incentives and information to invest in intangibles...' (BERR 2008: 13). The recognition by the British government and other governments that investment in intangibles is a key element of firm-based and national competitiveness is one of the drivers behind this book. Our argument is very simple: design matters and plays a crucial but often neglected role in the competitiveness of companies and nations.

Design has always played an important role in firm-based competitiveness. The relationship between design and competitiveness is complex, and it is also partially determined by geography. A product designed for the Swedish or German consumer may have limited appeal in other countries. This is a complex issue that is affected by national constructions of consumer behaviour. A good example is the introduction of automatic washing machines into North America. During the 1950s wringer manual washers outsold automatic washing machines in Canada, while in the United States automatic washing machines accounted for 75 per cent of sales. The explanation for these differences is complex but partly involves the fact that automatic washing machines required hot running water, as well as a water supply at a strong steady pressure. Many Canadian homes did not meet these infrastructural requirements. Infrastructural problems are only part of the explanation; another part concerns the perception amongst Canadian consumers that automatic washing machines used too much water and were too expensive compared to wringer machines (Parr 1999: 218–42). Parr's analysis of the consumption of domestic goods in Canada highlights many of the geographical issues that surround the relationship between design and competitiveness. National consumer behaviour and regulations are also important in this context. Thus, on 20 November 1940 Canada passed a regulation that forbade the introduction of new models of consumer appliances during the war. In contrast, American civil servants controlled output but encouraged companies to redesign products. This led to the design and development of new models that were lighter. It also meant that 'an American manufacturer who reduced the metal required to make a refrigerator from eighty-five to twenty pounds could increase production, remain within the controllers' quota, and keep his design team nimble in anticipation of the end of the war' (Parr 1999: 24).

This book provides an analysis of the relationship between industrial design and product, corporate or national competitiveness. This is a complex relationship, as industrial design can impact all parts of the production process of both goods and services. Industrial design's contribution to corporate competitiveness includes enhanced sales; greater profit margins compared to competitors' products; improvements in manufacturing efficiency, speed of repair, sales strategy, and design registrations and enhanced visibility for the firm and its products. Industrial design may influence and even determine revenues and costs. Design may impact revenues by increasing sales in existing markets, meeting preferences in other markets or by reducing production costs or product price. These impacts are associated with the creation of iconic 'designer' products, in improvements in performance and/or reliability and in the overall aesthetic appeal of a good or service.

This chapter provides an introduction to the relationship between industrial design, competitiveness, globalization and organizational strategy. The chapter is divided into three sections. The first section explores the shift from price- to design-based competitiveness or to an escalation in the design-intensity of products (physical goods and services) and production systems. It is perhaps important to differentiate between design as an outcome in terms of the creation of a 'designer product' and design as a continual process. The second section provides an overview of the ways in which firms incorporate design into their business strategies, while the third section provides an overview of the structure of this edited collection.

Design, competitiveness and inimitability

Design defined

Economists and policy makers often bifurcate the economy between processes of supply and demand. On the one hand, the supply side is where goods, services, companies and investment capital are oriented, whilst on the other hand, demand is where consumers' decisions are moulded. According to Koehn (2001: 3), most of the existing research on economic change has focused on understanding the changing dynamics of supply. This research has focused on understanding technological, financial and managerial innovations. To compensate for the emphasis placed on supply, Koehn provides a demand-side analysis that involves the construction of six case studies of firms that exploited available opportunities by 'comprehending consumers' emerging needs, developing a product that met these needs, communicating back and forth with customers, and actually delivering the product or service effectively' (Koehn 2001: 3). This is very much a story of marketing and brand creation or perhaps more correctly the process of 'market-making' in which a firm develops a new product and manages, at the same time, to create or manipulate consumer demand. Absent from this analysis is design (but see Lawrence and McAllister 2005: 107). This is perhaps unfortunate, as design plays a central role in the relationship between production and consumption. In an early analysis of the role design plays in manufacturing, Gloag suggests that archaeologists in the thirtieth century would soon realize that 'in the beginning [of the commercial machine age] there was industry without design ... [and that] the first industrial revolution – was practically over before design began to secure recognition as a technical operation' (1946: 15–16). In this account, designers create and modify products to meet the needs of consumers. The design process is a critical element in the link that exists between production and consumption or between supply and demand.

Design is a complex process that consists of many different but related processes. It can include the design of a good or service, but it also includes the design of packaging and graphic art. The relationship between industrial design and corporate competitiveness is complicated by the difficulty of developing a simple definition of industrial design. Industrial design can impact all parts of the production process of both goods and services. It can involve redesigning products to reduce production costs, by enhancing the efficiency of the production process, by reducing the quantity of materials incorporated into a product or by reducing installation or servicing costs (Rusten and Bryson 2007). The role of industrial design in product development includes ergonomics, aesthetics, ease of manufacture, product performance, efficient use of material, eco-friendly design and the relationship between the product and consumers. The complexity and all-embracing nature of the design process is further complicated by the different ways in which companies incorporate design into their value chain. In practice, the incorporation of design into a company will vary according to the nature of the firm's business activities and the design history and experience of the firm. A simple continuum can be identified between companies that develop and market products on the basis of design and the singular nature of their products and those that produce generic products or even copies of designer goods. This definitional difficulty means that it is perhaps best to adopt a general or abstract definition of industrial design. Ulrich and Pearson (1998: 352) and Gemser and Leenders (2001) define industrial design as the activity that transforms a set of product requirements into a configuration of materials, elements and components that comprise an artefact. Industrial design is part of the wider process of *product development*. This includes R&D activities undertaken by engineers that revolve around product creation, product testing and development and the interface with manufacturing production systems, as well as engagement with market research and marketing.

Designing inimitability

The competitiveness of firms is founded upon the efficient and effective manipulation of a series of inputs, including expertise, creativity, reputation and labour. Firms can compete on price, quality, speed, innovation, specialist products and services, brand and design. It can be argued that the 1950s were characterized by competition based on quantity, the 1970s by competition based on quality and the 1990s by competition based on sourcing and organization, and that the period from 2000 has been characterized by competition based on design (Perks et al. 2005: 112–3). This simple categorization of economic activity is questionable, but over the last 20 years there has been an escalation in the design-intensity of products. In recent years, a number of countries (for example, the Nordic countries, the UK, Korea and China) have begun to integrate design into their regional and national policies. The design budgets of American and European companies are growing by between 8 per cent and 20 per cent each year (Gemser and Leenders 2001: 28). According to one recent commentator, we are living through a period in which 'we now have an economy powered by human creativity...in virtually every industry, from automobiles to fashion, food products, and information technology itself, the winners in the long run are those who can create and keep creating' (Florida 2002: 5). Florida identifies the 'highest order of creative work as producing new forms or designs that are readily transferable and widely used – such as designing a product that can be widely made, sold and used' (2002: 69). Industrial designers are part of Florida's super-creative core; however, the process of industrial design and the role of industrial designers as key sources of competitiveness have been neglected, compared, for example, to the substantial literature on management consultancy (Bryson et al. 2004).

The emphasis that is now being placed on intangible investments like design is part of the response governments and companies have made to enhanced competition. This shift towards design-intensive production systems and design-intensive products represents a transformation in the policy environment, in the wider framework conditions that support innovation (Nonaka and Nishiguchi 2001) and in organizational strategy. In part, the shift towards design-intensive production systems is one outcome of the ongoing international division of labour in which manufacturing tasks are located in low-cost countries. This ongoing international division of labour does not imply that developed-market economies will no longer be involved in manufacturing. Two processes are at work (Bryson et al. 2008). First, companies develop blended production systems that are intended to capitalize on place-based differential competitive advantage (Bryson 2007: 34). Parts of a company's value chain remain in high-wage locations, and parts are distributed across low-wage locations. In effect, this is a spatial reordering of Chandler's (1962, Chandler et al. 1997) three pillars of corporate form – management, marketing and production. While management and marketing (including R&D and design) may remain in high-cost, developed economy locations, production may be shifted to low-wage locations. This strategy may also take issues of country risk into consideration to avoid overdependence on one national context. It may also be designed to enhance flexibility, with plants and offices located near to the market to enhance the ability of a firm to respond to client demands.

Second, companies located in high-cost locations can withdraw from price-based competition and compete on other variables related to design, brand and place (Rusten et al. 2007). Central to non-price-based competition, and drawing in part on the resource-based view (RBV) of the firm and the concept of 'inimitability' (Rugman and Verbeke

2002; Hoopes et al. 2003; Taylor 2006) are eight related strategies (Bryson et al. 2008) that firms can adopt:

1. Production of customized products that require close contact between producers and consumers.
2. Production of bundled products or products that hybridize products and services; for example, selling products with cradle-to-grave services attached (Bryson 2009a & b). This is especially important for machine tool manufacturers, as the profitability of their clients is partly determined by the reliability of the tools.
3. Development of skills and expertise that are difficult to replicate or transfer.
4. Creation of production processes that are difficult to replicate and often protected by patents and registered designs.
5. Design-based competition.
6. Continuous and ongoing process and product innovation.
7. Flexibility in responding to customers' requirements.
8. Delivery times and nearness to markets.

These are not mutually exclusive strategies and, depending on circumstances, firms can combine them in different ways. This list also suggests that by itself design-based competitiveness is not sufficient to ensure the survival and growth of a company. Design is part of a bundle of organizational variables and strategies that include, amongst other factors, organizsational strategy, the efficiency and flexibility of a firm's production systems, marketing, branding, access to finance, level of gearing and the quality of the firm's management team.

It is perhaps useful at this stage in the analysis to provide three examples of the relationship between design and inimitability. First, we shall consider the design of the iconic Coca-Cola bottle. The recipe for Coca-Cola was formulated in May 1886 and originally the syrup was delivered to a number of independent bottlers that obtained the right to fill bottles with a combination of syrup and soda water. The arrangement was that the syrup could only be supplied by the Coca-Cola Company. The independent bottlers used their own bottles which meant that Coca-Cola was delivered in bottles of all shapes and sizes. This made it extremely simple for imitators to provide copies of the company's product. Coca-Cola's brand name was insufficient protection and the company found itself involved in numerous lawsuits to protect its product. The solution to this problem was for Coca-Cola to design its own bottle. In 1915, the company held a competition for the design of a branded bottle which was won by the Root Glass Company. This was the company that created the iconic Coca-Cola bottle. A patent was registered for the bottle in 1915 (Albus et al. 2000: 28–9), and the design-centred intellectual property rights that Coca-Cola obtained with the new bottle enabled the company to protect its product from imitation.

Second, the success of the Apple Macintosh computer is related to the design of its operational system combined with the design of the hardware. In 1982, Steve Jobs, one of the founders of Apple, became aware of the work of Frogdesign, an innovative German design company run by Hartmut Esslinger. Steve Jobs invited Esslinger to present some ideas for a new family of Apple computers. Impressed by the designs, Jobs offered Frogdesign a $2-million-a-year retainer to enable the company to establish an office in the United States. Frog designed a small, beautifully crafted machine that was very different to other poorly designed ugly beige boxes (Sweet 1999: 22). The design ensured that 'not only could the dynamic nature and corporate identity of the company be successfully visualized, but the ensemble of monitor, keyboard and mouse became the creative ideal, above all for the culturally sensitized working processes of architects, graphic designers, product designers and advertising agencies' (Albus et al. 2000: 160).

Third, during the 1980s the Swiss watch industry was suffering from competition from the Japanese watch industry. One response to this competition led to the development of the Swatch watch. The watch mechanism was redesigned to reduce the number of components from over 100 to no more that 54 (Albus et al. 2000: 158). The assembly was completely automated and the components were fixed together using ultrasonics rather than screws, with the result that the Swatch could not be repaired. Every Swatch mechanism is exactly the same, and only the watch face and the strap offer any possibilities for design. Every year the Swatch Design Lab in Milan designs around 70 new versions of the Swatch. Through design, the Swatch transformed the watch into a fashion accessory; a high-quality, collectable but still mass-produced product (Glasmeier 2000: 249).

These three examples illustrate some of the many meanings that are associated with design as an inimitability strategy. The example of Coca-Cola highlights the role design plays in the packaging and display of a product, whilst the Apple example concerns the enhancement of a product's functionality and appearance. The Swatch example demonstrates the ways in which design can transform a product into a design-intensive fashion accessory. All three examples make it difficult for other companies to copy these strategies, as intellectual property protection exists to deter imitators and regular redesigns make it difficult for imitators to mimic a firm's current product range. We now turn our attention to understanding the role that design plays in enhancing the competitiveness of firms.

Design

The current financial crisis and economic downturn means that it is even more important for companies to find clever ways of surviving and winning market share. A new invention or the development of a product with an appealing design may convince the market and investors that a company is still financially viable and potentially profitable. Intelligent design can reduce production times, reduce the raw materials required to produce a product, enhance the sustainability of a product, reduce the amount of warehousing required and even reduce capital inputs. During an economic downturn, internal capacity within a firm may be available that could be usefully assigned to enhance the design of existing products or to create new products. Products coupled with a place-based identity may further ground companies in a particular country and the origin of a product to a specific place. A strong design identity coupled with sophisticated industrial knowledge and high-value products might even make it possible for a firm to continue to manufacture in high-cost locations.

The design of a product plays an important role in branding and consumer recognition. A product may have a recognizable shape so that a consumer will immediately understand how it can be used and also relate the product to a particular company or even designer. Design-centred product recognition is in fact a shortcut to making a decision to purchase a product. In some cases, the design functions can personify the product so that the consumer is aware of the designer's or the manufacturer's reputation. This identity may also directly or indirectly be linked to geography, for example a Finnish product or Canadian designer. Industrial design can also become a type of creative knowledge that is brought into practise when a product's design and image or identity becomes part of the image of the end consumer. In this case, 'you are what you wear or consume'. The name of the brand or the designer may become more important than the actual product.

An increase in the design-intensity of products enhances the differences that exist between designer products and other products. It will, at the same time, present design-intensive companies with the challenge of ensuring that the competitive advantage that they have created through design is maintained and even enhanced (Hall 1993). Consumers expect that expensive high-quality products will also be designer products. This means that the industrial designer often faces the challenge of updating a product or creating a new product but, at the same time, retaining aspects of the product that resonate with existing consumer collective understanding and experience of this product type (Farstad 2003).

Design is a strategic tool for industrial competition that enables a firm to offer something different and perhaps better than its competitors (Trueman and Jobber 1998). This also means that designers need to be able to demonstrate the ways in which design is associated with the development of successful products and enhanced company performance. Designers need to be trained to better understand different industries and their specific needs, and they must also be able to communicate these ideas in a way that arouses consumer enthusiasm for a product. A product is usually created or modified through the action of multidisciplinary project teams. It is desirable that industrial designers are included into these product terms from the start, rather than design being a process that is added to a product after the R&D phase has been completed. Bringing designers into the boardroom and making use of multidisciplinary design teams represents one way of increasing the use of industrial design knowledge in the production process. It is especially important to bring management, technicians and marketing people together to work on design projects.

A designer can work alone or with other designers, and many design firms are extremely small. Working together as part of a project team is an advantage for design projects for several reasons. The involvement of more than one designer in a project reduces the risks associated with illness and capacity problems. There are positive social benefits from working as part of a team that include increased capacity, but also the ability to develop ideas through discussion. It is also advantageous for a designer to follow a design through its various stages, from the formulation of a prototype through to marketing and sales. Firms intending to export a product find it useful to seek the advice of designers experienced in the preferences and cultural differences of consumers located in the primary export markets. In this instance, the industrial designer can act as a broker between cultures and the product development or modification process. Engineers involved in R&D must be educated to appreciate the benefits that come from design expertise, and designers must also be trained to understand the engineering professions. The development of clear theoretical frameworks and practices that can be communicated through formal teaching, research and dissemination are very important in reaching these goals (Rusten 2008).

Design involves much more than working on the shape and appearance of products. Design also involves the introduction of products that may set an industry standard for a particular style. These are well known motivations in the textile, garment, furniture, automotive and electronic industries. A design might involve the ability of a product to meet standards and regulations that may concern industrial specifications, safety, environmental or social issues. Many public sector infrastructural developments require that universal design be applied so that the development is designed for all citizens and that everyone has the same equal opportunity to participate fully in society and working life (Norwegian Directorate of Health 2008). It might also be about being able to introduce more efficient production processes, or better quality and more environmentally friendly products than competitors. In this case, a company can deliberately design products targeted at sustainable consumption or the eco- or green consumer. One element of an active design strategy might involve the acquisition of a patent or design registration to protect the property rights of a firm.

There are many different types of designers and design processes, and this heterogeneity is reflected in the difficulty of producing a simple definition of design. The design process includes fashion designers as well as furniture designers or even interior designers. Other designers use titles related to the type of activity they are involved with, for example, graphic designer, product designer or industrial designer, or just 'designer'. Some designers have a pure art educational background, others are more technologically oriented and some have a combination of these two traditions. In some cases, designers are employed on full-time contracts, and in other cases, they are employed on short-term contracts on a product-by-product basis. Combinations of these two types of employment relationship also exist even within the same firm. A famous design star may be employed on a short-term contract, while being supported by a firm's full-time designers. There are also examples of firms that have purchased the rights to a designer product that has not yet been commercialized. A designer can also contact a manufacturer to suggest that an item should be developed for production and, if successful, may follow the product all the way through the design, development and production process until the product reaches the market. On a much larger scale, companies like IKEA are doing something similar, but are also responsible for the sale of the products.

Neither should we forget the cultural and geographical dimensions of design. Bucciarelli (1994) describes design as being an international language, but a product that is successful in one country might not be considered to be saleable in another. A company, therefore, needs to understand the cultural nuances and preferences that vary from country to country (Bucciarelli 1994). Designers that are focused on the user context must undertake fieldwork to collect sufficient information about consumer behaviour. This is done in the preliminary stages of a design project, as well as researching user experiences at a later stage when the product has been introduced into the marketplace. Exploring consumer use of an existing product is extremely valuable in terms of introducing product improvements or for the design of new products. In this context, designers are engaged in fieldwork that resembles the participant observation methods that are well known amongst anthropologists. For example, the Norwegian designer Einar Hareide followed an ambulance boat to observe working conditions and practises amongst the crew and the activities of the medical team whilst they worked on patients. An important aspect of designers' research process involves interviewing existing users and consumers about their use and perception of an existing product and also obtaining suggestions for product improvements (Rusten and Bryson 2007).

Designers must remain up-to-date in relation to different technologies, standards, styles and trends (Rusten 2008). Identifying ideas for product or process improvements from the shop floor is another way in which designers can identify new ideas that might result in the formulation of a new product. Designers will often be more focused on user requirements and consumer voices or reactions to products than engineers. This difference in the competencies and skill sets of engineers compared to designers and also the different methodological approaches deployed by these professions suggests that product development teams must be multidisciplinary.

The broad range of perspectives and the variety of design cases that are presented in this book demonstrate the importance and significance of design expertise and knowledge as an essential bridge between the requirements of users and the professions involved in the technological development of goods and services. Much of this design work has brought the voices of customers and clients to the forefront of the product development process. Being very much aware of the needs and preferences of potential consumers or users is now considered an essential element of the product development process by many firms. In contrast, during the early years of the Industrial Revolution, most products were just produced and sent directly out into the marketplace. Since the 1920s the social status of the designer has been transformed from an activity that was almost invisible into a formalized profession.

Design has become more than just a pure business concern. Design has become part of the policy toolkit as governments have begun to consider it as a new and powerful knowledge component of many products and services. A regional policy that includes a focus on design can be a way of creating more categories of jobs as well as strengthening the competitive position and geographical associations of businesses involved in production as well as in other sectors. Joint seminars, scholarships and the use of research investment funds, public arenas and public events are all tasks that may encourage the development of new design projects that engage a whole range of businesses. Design is considered one way of ensuring that some manufacturing production is retained in countries with high labour costs.

The emphasis that some governments are placing on design has many different drivers. First, design is considered a useful tool for enhancing the added value that can be incorporated into a product. Second, design can be used to transform products into inclusive products that can be used by all people. Third, design can enhance the sustainability of a product by increasing energy efficiency and reducing the raw materials used in the production process. It can also simplify the recycling process. Fourth, design is also considered one way of enhancing a region or nation's reputation for the production of high value-added designer products that contain national or regional associations or symbols. Fifth, designers along with dancers, musicians, filmmakers and other artists are part of the creative economy and this group of occupations is considered to be important for the vitality of regional economies.

The structure of the book

This book is divided into two parts. The first part explores the relationship between industrial design and national competitiveness, and the second part explores design and the competitiveness of firms.

The first part of the book begins with Chapter 2 and an analysis by Bryson of the relationship between national competitiveness and the emergence of design-centred economic policy. Design has always played an important role in the competitiveness of some firms, and this role has been recognized by governments for over 200 years. The analysis begins by exploring the relationship between design and the development of a division of labour, before exploring the development of design-centred national policies. The education of consumers, manufacturers and designers has played a central role in the formulation of design-centred policies, and the educational focus has been supported by a series of major international exhibitions and competitions. A central part of the argument is that companies compete in many ways: on the basis of price, process, product, innovation, patents, brands, fashion and design. These different forms of competitiveness are mirrored in Chandler's three pillars of corporate form – management, marketing and production. These represent generic business skills that are applicable to the majority of firms (Chandler 1977). Nevertheless, Bryson argues that design should be considered an additional fourth pillar of corporate form.

In Chapter 3, Solitander explores the relationship between design and national competitiveness through an analysis of the emergence of a creative economy in Finland. After World War II, Finland successfully developed a new identity constructed around design and architecture and through the export of cultural products. This chapter explores this transformation and in particular the development of a discourse of Finnish design. On an international scale, Finnish-designed products struck a chord with consumers in the 1950s and 1960s, as they mirrored the spirit of the times and the changes in Western society – for example women's equality, urban migration and modern lifestyles. In many respects, however, the impact of Finnish design was perhaps more important on an imagined level. One of the strengths of Solitander's analysis is the identification of tensions or contradictions in discourses that emphasize design and creativity in national policy agendas. Some of these include the tension that exists between the consumerist lifestyle of the creative class and issues of ecological and social sustainability. Another issue concerns uneven development and the silencing of other voices and arguments in economic policies that are dominating debates concerning the activities of creative industries and occupations. Part of Solitander's argument involves an analysis of Arabianranta – the City of Art and Design, Helsinki. The aim of this project was to establish Arabianranta as the leading centre of art and design in the Baltic context. This is an interesting policy-driven initiative that plays a central role in Helsinki's overall innovation strategy and is a superb example of the ways in which contemporary policy is being used to enhance the design-intensity of a developed market economy.

In Chapter 4, MacPherson and Vanchan present an analysis of two surveys that were undertaken in the United States in 2005 and 2006. The first survey was based on a sample of 85 design consultancies, and the second survey was based on a sample of 68 large US producers of durable goods. This chapter addresses a number of themes that have recently emerged in the literature on advanced producer services. First, specialist design firms tend to compete on the basis of reputation, service quality, referrals and trust-intensive relationships with longstanding clients. The available evidence suggests that cost-driven externalization has never been a significant factor in the use of external designers by client firms. Instead, clients typically seek new or innovative services that cannot easily be replicated via in-house effort. A second theme is that design contracts have become increasingly performance-linked (that is, risk-sharing agreements are common). Third, the industry is gradually moving toward technological forecasting and design tradeoff analysis. Design tradeoff analysis (DTA) allows clients to statistically evaluate potentially competing design objectives such as product performance versus ease of manufacture. Fourth, some of the more successful firms in this industry belong to elaborate innovation networks that connect different types of specialists. The presence of these networks explains why so many micro-businesses in this sector can offer a full range of design services.

Rantisi continues the North American focus in Chapter 5 with an analysis of independent fashion designers and high-end apparel manufacturers in Montréal, Canada. 'Independent designers' here refers to firms in which the designer is both owner and creative director, and 'high-end apparel manufacturers' refers to firms with in-house design capability, where the designer or design team provide creative direction and are often employees. Rantisi's analysis is based on over 40 interviews with key industry actors, including 28 apparel producers. The primary research question concerns the ways in which designers and manufacturers obtain and transmit key market information, highlighting the role that cultural intermediaries play in facilitating this process. The analysis suggests that while Montréal designers have forged close relations with local institutions and have been successful in developing popular designs for the local market, relations to key international institutions are lacking. As the Montréal industry becomes increasingly globalized, with a heavy reliance on US exports, the lack of such linkages hinders designers' ability to 'embed' themselves in non-local markets and develop competitive designs.

In Chapter 6, Vinodrai explores the relationship between design and national competitiveness through an analysis of design work in Canada. Business and policy discourse increasingly identifies design as critical to competitiveness and innovation. This resonates with an emerging literature that suggests that 'creativity' is crucial to the economic performance of firms, industries, and city-regions. Research investigating the creative and cultural industries reveals that the underlying labour markets that support this work are often highly localized and project-based within urban settings. Within these urban labour markets, local social dynamics are important in mediating outsourcing via freelancing and other contractual arrangements. Vinodrai's analysis explores the geography of design work in Canada and offers insight into the complex set of local and non-local outsourcing relationships that shape design work. The chapter reveals the embeddedness of design knowledge and labour markets, as well as the importance of proximity to clients and markets as important for reinforcing a local geography of design work. Yet, tensions arise from the need to balance proximity to clients and markets with proximity to production facilities. This is especially relevant because design-related production in some industries increasingly takes place in 'offshore' locations (e.g., China). Overall, the analysis reveals that the actual work of design in Canada remains concentrated in only a handful of urban places, and this geography is reproduced though a complex set of institutional and social dynamics.

The final chapter in this part of the book is by Hutton. In Chapter 7, Hutton provides a detailed analysis of design in cities with a particular emphasis placed on the saliency of place as well as space, and acknowledging the mix of development factors and interdependencies influencing location. Hutton develops a model of design activity at the urban-regional scale which acknowledges the logics of industrial location, as well as more recent efforts to depict the spatiality of cultural production in the metropolis by giving greater weight to social, environmental and policy factors. A set of illustrative examples of key sectors and instructive case studies at the city and district level demonstrates the rich empirical dimensions of design activity in the city, informed by this earlier conceptual enterprise and by an extensive program of field work, and including, primarily, sketches of design industry formation in two cities: London and Vancouver. The discussion includes a description of the tensions produced by cost pressures and burgeoning outsourcing tendencies, the development of more extended production networks and global recruitment of key labour cohorts, which modify the operating characteristics of design industries within localized clusters. To a large extent, the geography of design activity has taken the form of industry clusters situated within the central and inner city, as exhibited in the London and Vancouver cases. But a more nuanced perspective discloses some important variants and departures from the classic clustering model, as in the engagement of design firms within more extended regional production networks, in subcontracting arrangements and in outsourcing. Over time these features of the geography of design are likely to increase in importance, following tendencies in other production regimes and sectors over the last three decades.

In the second part of the book the focus of the analysis shifts towards exploring the relationship between design- and firm-based competitiveness. This part begins with Chapter 8, written by Hansson et al., which explores the relationship between design gender and competitiveness. Firms compete on price and quality, but can also compete on inclusion, image and sustainability, and this can include the strategic deployment of gender and design as a competitive strategy. This chapter begins by exploring debates surrounding the commercialization of gender by building upon some of Sparke's (1995) earlier work on pink design. The chapter summarizes the findings of a 2-year multidisciplinary research project on gender, design and business organization that combined expertise from culture studies, marketing and design. On the one hand, the analysis revolves around market segmentation, but on the other hand, detailed case studies of innovative design projects intended to challenge gender stereotypes are described and analyzed. In the chapter, the authors survey a variety of consumer goods with gender implications and develop a typology to account for the ways in which companies compete on the basis of gendered designs. In the final section of the chapter, a detailed analysis of Volvo's YCC concept car is undertaken. This case study explores the formation and organization of a design team that was created to develop a car designed by and for women. This is an interesting case study, as it highlights tensions between the design team and company policy and also shows the ways in which gender-inspired innovations were eventually transformed into relatively gender-neutral innovations.

In Chapter 9 Rusten provides an analysis of design within the Norwegian maritime industry in which she describes situations of co-creation between designers and clients. Research into design-intensive products has tended to concentrate on the analysis of the fashion industries, for example, the furniture or automotive sector, and has largely ignored sectors like the maritime industry. The aim in this chapter is to identify and explore the business models, organizations, products and geographies related to industry-oriented design projects that are aligned with the maritime sector. The analysis also highlights the importance of the transfer of concepts and products between different industrial sectors. Industrial designers working with different categories of products and clients may be able to transfer designs between industries, and in doing so increase their profitability.

In Chapter 10, Gabrielsen et al. set themselves an important question to investigate: is design good business? The chapter is based on the analysis of data compiled on 25 of Denmark's largest companies. The results of the analysis are reassuring for a book on the relationship between design and competitiveness. The findings indicate that the main driver of company performance is product design. This is an important finding that supports research commissioned by public sector organizations such as the Design Council (London) (Design Council 2004, 2005). This does not mean that all design is good business, as it is possible for design-intensive firms to produce designer products that fail in the market. Rather, this study acknowledges that companies investing in design tend to outperform companies that fail to invest in design on a series of measures of corporate performance.

Overall, this book is a call for academics and policymakers to consider the relationship between design and company performance as an important component in firm- and national-based competitiveness. It is possible to argue that the competitive strength of nations is founded upon intangible investments in not only education and training, but also design. The arguments presented in this book illustrate that design matters in multiple ways and that an important research agenda still needs to be developed to explore the relationship between design and competitiveness.

References

Albus, V., Kras, R. and Woodham, J.M. (Eds) 2000. *Icons of Design: The 20th Century*. Prestel, Munich.

BERR. 2008. *Manufacturing New Challenges, New Opportunities*. BERR, London.

Bryson, J.R. 2007. 'A "second" global shift? The offshoring or global sourcing of corporate services and the rise of distanciated emotional labour'. *Geografiska Annaler* 89B(S1): 31–43.

Bryson, J.R. 2008. 'Value chains or commodity chains as production projects and tasks: Towards a simple theory of production', in Spath, D. and Ganz, W. (Eds), *The Future of Services: Trends and Perspectives*. Carl Hanser Verlag, Munich, 265–84.

Bryson, J.R. 2009a. 'Service innovation and manufacturing innovation: Bundling and blending services and products in hybrid production systems to produce hybrid products'. in Gallouj, F. (Ed.), *Handbook on Innovation in Services*. Edward Elgar, Cheltenham, in press.

Bryson, J.R. 2009b. *Hybrid Manufacturing Systems and Hybrid Products: Services, Production and Industrialisation*. University of Aachen.

Bryson, J.R., Daniels, P.W. and Warf, B. 2004. *Service Worlds: People, Organizations, Technologies*. Routledge, London.
Bryson, J.R., Taylor, M. and Cooper, R. 2008. 'Competing by design, specialisation and customization: Manufacturing locks in the West Midlands (UK)',. *Geografiska Annaler* 90(2), 173–86.
Bucciarelli, L. 1994. *Designing Engineers*. MIT Press, Cambridge MA.
Chandler, A. 1962. *Structure and Strategy*. Harvard University Press, Cambridge MA.
Chandler, A., Amaroti, F. and Hikino, T. 1997. 'Historical and comparative contours of big business', in Chandler, A., Amaroti, F. and Hikino, T. (Eds), *Big Business and the Wealth of Nations*.Cambridge University Press, Cambridge MA, 3–23.
Chandler, A.D. 1977. *The Visible Hand: The Managerial Revolution in American Business*, Harvard University Press, Cambridge, MA.
Daniels, P.W. and Bryson, J.R. 2002. 'Manufacturing services and servicing manufacturing: Changing forms of production in advanced capitalist economies'. *Urban Studies* 39(5–6), 977–91.
Deitz, R. and Orr, J. 2006. 'A leaner, more skilled U.S. manufacturing workforce'. *Current Issues in Economics and Finance* 12(2): February/March, 1–7.
Design Council. 2004. *The Impact of Design on Stock Market Performance: An Analysis of UK Quoted Companies 1994–2003*. Design Council,.London.
Design Council. 2005. *Design Index: The Impact of Design on Stock Market Performance*. Design Council, London.
Farstad, P. 2003. *Industridesign*. Universitetsforlaget.Oslo.
Florida, R. 2002. *The Rise of the Creative Class and How It's Transforming Work, Leisure, Community and Everyday Life*. Basic Books, New York.
Gemser, G. and Leenders, M. 2001. 'How integrating industrial design in the product development process impacts on company performance'. *The Journal of Innovation Management* 18, 28–38.
Glasmeier, A.K. 2000. *Manufacturing Time: Global Competition in the Watch Industry, 1795–2000*. Guilford, New York.
Gloag, J. 1946. *Industrial Art Explained*. George Allen & Unwin, London
Hall, J. 1993. 'Brand development: How design can add value'. *Journal of Brand Management* 1:2, 94–100.
Hoopes, D., Madsen, T. and Walker, G. 2003. Guest editors' introduction to the special issue: 'Why is there a resource-based view? Towards a theory of competitive heterogeneity'. *Strategic Management Journal* 24, 889–902.
Koehn, N.F. 2001. *Brand New: How Entrepreneurs Earned Consumers' Trust from Wedgwood to Dell*. Harvard Business School Press, Boston
Lawrence, P. and McAllister, L. 2005. 'Marketing meets design: Core necessities for successful new product development'. *The Journal of Product Innovation Management* 22, 107–8.
Nonaka, I. and Nishiguchi, T. (Eds) 2001. *Knowledge Emergence: Social, Technical, and Evolutionary Dimensions of Knowledge Creation*. Oxford University Press, Oxford.
Norwegian Directorate of Health. 2008. *Who Needs Universal Design. Universal Design in Public Sector*. Helsedirektoratet-Deltasenteret, Oslo.
Parr, P. 1999. *Domestic Goods: The Material, the Moral, and the Economic in the Postwar Years*. University of Toronto Press, Toronto.

Perks, H., Cooper, R. and Jones, C. 2005. 'Characterizing the role of design in new product development: An empirically derived taxonomy'. *The Journal of Product Innovation Management* 22, 111–27.

Rugman, A. and Verbeke, A. 2002. 'Edith Penrose's contribution to the resource-based view of strategic management'. *Strategic Management Journal* 23, 769–80.

Rusten, G. 2008. 'Designtjenester og geografi (Designservices and geography)' in Isaksen, A., Karlsen, A. and Sæther, B. (Eds), *Innovasjoner i norske næringer – et geografisk perspektiv*. Bergen: Fagbokforlaget, 223–41.

Rusten, G. and Bryson, J.R. 2007. 'The production and consumption of industrial design expertise by small- and medium-sized firms: some evidence from Norway'. *Geografiska Annaler* 89B(S1), 75–87.

Rusten, G., Bryson, J.R. and Aarflot, U. 2007. 'Places through products and product through places: Industrial design and spatial symbols as sources of competitiveness'. *Norwegian Journal of Geography* 61, 133–44.

Rusten, G. and Stensheim, I. 2007. 'Teknologiutvikling og design blant leverandører til oppdrettsnæringen' in Aarset, B. and Rusten, G. (Eds), *Havbruk; Akvakultur på norsk*. Fagbokforlaget, Bergen.

Sparke, P. 1995. *As Long as It's Pink: The Sexual Politics of Taste*. Harper Collins, London.

Sweet, F. 1999. *Frog: Form Follows Emotion*. Watson-Guptill, New York.

Taylor, M. 2006. 'Fragments and gaps: Exploring the theory of the firm' in Taylor, M. and Oinas, P. (Eds), *Understanding the Firm: Spatial and Organizational Dimensions*. Oxford University Press, Oxford.

Trueman, M. and Jobber, D. 1998. 'Competing through design'. *Long Range Planning* 31:4, 594–605.

Ulrich, K.T. and Pearson, S. 1998. 'Assessing the importance of design through product archaeology'. *Management Science* 44:3, 352–369.

I
Industrial Design and National Competitiveness

2
Industrial Design, National Competitiveness and the Emergence of Design-Centred Economic Policy

John R. Bryson

> In early stages of manufactures, it is mechanical fitness that is the object of competition: as society advances it is necessary to combine elegance with fitness; and those who cannot see this must send their wares to the ruder markets of the world, and resign the great marts of commerce to those of superior taste who deserve a higher reward.
>
> (Wornum: [1851], 1970: I)

A national economy consists of a complex array of private and public sector organizations supported by physical infrastructure and a set of policy and legal frameworks and legislation. The infrastructure and policy frameworks are intended to support and encourage wealth creation and have developed or evolved over a long time period. Government policy is created at a variety of spatial scales to regulate, support and enhance wealth creation. The policy environment that supports and also regulates private sector activities is reactive and proactive: proactive as it attempts to lay down the conditions for wealth creation and reactive as it responds to events, for example, company closure, recession or enhanced competition. In many instances policy is developed that is built upon fashionable theories popularized by consultancy companies and gurus (Bryson 2000). There are many drivers behind the formulation of a policy response: fashion, peer pressure, copying, response to a crisis and the vision of an individual or group of individuals. In most cases, policy formulation is intended to address a particular problem or is driven by a political agenda that might be informed by the latest policy fashion.

The national policy environment is dynamic as elected politicians and their officials attempt to respond to current circumstances and pressures; one of the political pressures is the requirement for policy to be translated into identifiable and visible actions and actions that will contribute towards the re-election of a politician or political party. On the one hand, the policymaking community is influenced by the latest policy fashions, for example, the emphasis that is currently being placed on the creation of green collar jobs, or jobs in the creative industries. On the other hand, policymakers must respond to local circumstances and be willing to ignore pressures that might be placed upon them to adopt the latest policy fashion. There is a real danger that regional policymakers adopt national and international fashionable policy at the expense of crafting locally sensitive and nuanced policy frameworks. One recent report on the impact of the current recession on the UK has gone as far as arguing that:

> Trying to grow high-technology clusters from scratch is unlikely to succeed. But the UK's regions and nations do have unique strengths that can form the basis for growth. Places do not just vary in their innovation performance: they vary in how they innovate, using different combinations of local assets and resources to better or worse effect. Different locations have different sources of innovative strength…. These strengths are what regional economic strategies should be based on. To date, we have seen too much imitation in UK regional economic policy. For example, eight of England's nine regional economic strategies prioritise biotechnology or health sciences. (NESTA 2009: 8)

In the UK, regional development agencies have adopted clusters as a key element of their policy toolkit (Bryson et al. 2008a). This has perhaps been unfortunate, as it has led to a proliferation of policies that have been designed to support proto-clusters and clusters that resonate with fashionable policy, for example, high technology, creative industries or health technology clusters. All this is a 'clear example of good aspirations running too far ahead of reality' (NESTA 2009: 11). A regional policy must be crafted to meet the needs of a local economy. This means that not all places should compete on the basis of high technology or the creative industries, but rather policy should enhance the competitiveness of a region's existing firms, encourage the establishment of new indigenous firms and also attract foreign direct investment. However, the resilience of a regional economy depends in part on the diversity of its industrial structure. Concentration on one industrial activity leads to dependency and also exposure to economic downturns and competition from companies located elsewhere. It is essential that a distinction be made between policy that is targeted at a particular industry and more generic economic policies. This chapter explores the evolution of one type of generic policy intervention, the development of national and regional policies that have been formulated to try to encourage firms to compete on design as well as price (Bryson et al. 2008b).

Companies can compete in many ways: on the basis of price, process, product, innovation, patents, brands, fashion and design. These different forms of competitiveness are mirrored in Chandler's three pillars of corporate form – management, marketing and production. These represent generic business skills that are applicable to the majority of firms (Chandler 1977: 419). Nevertheless, I argue that design should be considered as a fourth pillar of corporate form. These four pillars require policy formulation that is generic as well as sector-specific. In the case of design, consumers and the general public need to be educated to appreciate good design. On the one hand, generic design education should be an integral part of the educational system, whilst on the other hand, design should be an integral component of sector-based economic policies.

This chapter explores the ways in which national states have developed and introduced regional and national policies to encourage firms to enhance the design-intensity of their products and practices. The chapter is divided into five sections. The first section explores the application of a division of labour to manufacturing during the early years of the Industrial Revolution and the subsequent development of design-based competition. This section also explores Chandler's three pillars of corporate form and develops the argument for the existence of a fourth design pillar. The second section explores the relationship between design and the industrial revolution by exploring the ways in which design was introduced as a key competitive strength by a number of companies. In the third section the ways in which industrial design was incorporated into national economic policy in the UK is examined whilst the fourth section examines the rise and role of major industrial design exhibitions from the early years of the nineteenth century into the twentieth century. The fifth and final section presents a conclusion to the argument.

From the division of labour to design-based competition

Prior to the Industrial Revolution, the production of products was dominated by a system of craft production in which individual craft workers controlled design and production. Craft workers were trained through apprenticeships, and standard designs were handed down from generation to generation. The transition towards an industrial society is also a transition away from a society dominated by craft workers. It is possible to identify three phases in this transition towards industrial production. First, there was a period when everything that was made was essentially a craft object and everything during this period was essentially handmade. Second, during the Renaissance a distinction developed between the concept of craft and fine art with the latter being considered superior. Third, the Industrial Revolution is associated with a separation between craft object and industrial products or things made by machines (Lucie-Smith 1981: 13). The Industrial Revolution is sometimes conceptualized as a conflict between craft and industrial production; craft production is associated with individuality and quality and industrial production with quantity and uniformity. The transition from a craft- to industrial-dominated economy began gradually with the substitution of machines for hand tools, the development of primitive factory systems and perhaps most importantly with the development of a division of labour. Craft production increasingly became incorporated into factory production systems, and an industrial economy driven by competition began to develop.

During the Industrial Revolution, two important processes were at work: the division of labour and competition. We will explore each of these processes in turn. An ongoing division of labour based on specialisation of tasks occurred in all industries and in the case of pin manufacturing was documented by Adam Smith in 1776 (Smith 1977). Initially, the introduction of a division of labour can be understood as one way in which employers tried to replace the individuality that was at the core of craft work with uniformity of process and product. A good example of this process is found in the case of Josiah Wedgwood and the 'Etruria' porcelain factory that he established in Stoke-on-Trent in 1769. At the beginning of the eighteenth century, most pottery was produced for the local market. By 1750, some potters had begun to export to other regions of England as they had developed expertise in particular types of pottery. When Wedgwood established his workshop, most potters sold their products by sending completed pots directly to market or to a merchant. In 1774, Wedgwood decided to alter this practice by producing a catalogue. Catalogue selling only works when factories can produce goods, uniform in quality as well as detail. Wedgwood's problem was that potters liked to show their skill and creativity rather than produce standardized pots. To prevent product variation, Wedgwood introduced enamel-printed transfers that were mass-produced and applied to the pots before firing. The hand painters could no longer stray from the designs printed in the catalogues.

The only other source of variation came from the potters who were still able to introduce variations into the form of a pot. In the 1750s, to overcome this problem, Wedgwood divided the work required to manufacture a pot into seven different occupations (Forty 1995: 32). Each employee undertook a single task, and because no employee was responsible for a single pot, no individual was able to alter the design. This was Wedgwood's attempt to convert his employees into 'such machines of men [sic] as cannot err' (Wedgwood, cited in Finer and Savage 1965: 82). At each stage of the production process, however, employees were still able to make small variations to the design. Wedgwood kept dividing the production process into more and more specialist tasks. This increased his control of the process, and it also permitted him to replace skilled labour with less skilled, cheaper labour. Josiah Wedgwood eventually was able to standardize the production of pots by introducing a relatively sophisticated division of labour into his factory. The introduction of a division of labour also transferred responsibility for the design process away from the craft workers to management. Wedgwood was able to develop a distinctive design identity for the products produced by his factory, and the development of design-intensive products became one of the company's primary sources of competitive advantage.

The introduction of a division of labour transferred control of the production process from the shop floor workers to the managers. It also reduced production costs, enhanced the efficiency of production processes and increased output. The English watch industry prior to the introduction of factory production was based upon a system of small batch production. This meant that even the most famous watchmakers did not make all the components that went into one of their watches. Thus, one of the greatest English clockmakers of the seventeenth century, Thomas Tompion (1639–1713), relied on a complex division of labour. This division of labour was described by Sir William Petty in the following way:

> In the making of a Watch, if one man shall make the *Wheels*, another the *Spring*, another shall engrave the Dial-plate, and another shall make the *Cases*, then the *Watch* will be better and cheaper than if the whole work be put upon any one man.
> (Petty, W. [1678], 2004)

According to Adam Smith, the increase in the quantity of work that results from the application of a division of labour to a production process is the product of three factors (Smith 1977: 112–3). First, the dexterity of employees improves and this contributes to an increase in speed. Second, time is lost when an employee has to change from one type of work to another. Tools have to be found, the working area must be altered or the worker must move to a different part of the shop floor. Time is also lost as the worker adjusts to a new task. Third, labour can be replaced by machines, permitting cheaper and faster production. A division of labour precedes the replacement of a task by a machine. It is through a division of labour that distinct parts of the production process become visible, and as they become visible machines can be developed to replace labour. Ultimately, the division of labour utilized by Tompion to create watches led to the demise of the English watch industry. During the middle of the nineteenth century, Pierre Frédéric Ingold developed a process for making watches using precision machines to manufacture the components. This made it possible to mass-produce watches to a standard design using relatively unskilled workers. This new technology was adopted by the American, Swiss and French manufacturers, but not by the English makers. The result was that the English watch industry was destroyed by the import of millions of cheaply made Swiss and American watches (Thompson 2008: 12). The application of design to a production process enables a common production platform to be developed (Taylor and Bryson 2006; Bryson et al. 2008b). Many varieties of a product can be produced that draw upon a common set of parts. Design-intensive firms can exploit the advantages that come from mass production, but are also able to create mass-produced customized products.

The demise of the English watch industry brings us to the second process, that of enhanced competition. The rise of mass production is associated with the rise of mass consumption and also a series of new occupations that bridged the gap between production and consumption. These new occupations were advertising, marketing and design. The rise of these new occupations is another example of the division of labour that led to the development of a host of specialist marketing functions that were an integral part of the mass production of things. The Industrial Revolution is associated with the growth of trade and the establishment of European colonies in Asia, Africa and the Americas. The new colonies provided the raw materials and, at the same time, were considered to be captive markets. But it was not this simple. Europe had been an important market for products produced in India and the Far

East ever since the establishment of trading routes, and these imports intensified with the establishment of reliable shipping routes in the seventeenth century. The East India Company played an important role in this trade, and the trade was driven by goods that could not be produced in Europe, but also by price-based competition. As in the current century, labour costs were much cheaper in the Far East than in Europe, and European companies could only respond by developing the division of labour and by introducing mechanisation. Far Eastern goods also influenced the design of European products, including porcelain, wallpaper and furniture. Thomas Chippendale, the English eighteenth century cabinetmaker, was interested in Chinese decorations and was stimulated by imports of lacquer cabinets and screens, fabrics and porcelain. He developed a style that came to be known as Chinese Chippendale which included carved lattice-work (Gloag 1949: 118).

The rise of mass production is associated with the development of companies that competed on price, the organisation of production (Bryson and Rusten 2008) and on design. In many cases the emphasis in the literature has been placed on price-based competition, mechanisation and deskilling (Braverman 1974), and on the organisation of production systems (Dicken 2003), and it is possible to argue that insufficient attention has been given to design-based competition. These various strategies are not mutually exclusive (Bryson and Rusten 2008). Good design can reduce production costs and can support price-based competition. With the rise of factory capitalism, combined with a developing transportation industry, companies increasingly had to consider their position in the market place. Position, in this instance, refers to the ability of a firm to differentiate itself from its competitors. Differentiation could be on the basis of price, quality of product, location, services attached to products, established brand, design, product range and the development of niche products. Initially competition tended to be price-based, but this was supplemented by the establishment of strong brands and also by design. The comparative weakness of intellectual property protection and especially the absence of an effective international patent system meant that companies had to continually innovate to maintain their position in the marketplace. In his history of American business, Chandler explores this shift towards continuous innovation (1977) by exploring the activities of two companies: Singer Sewing Machines and Eastman Kodak. American patents rarely provided protection in overseas markets, and rarely did one company control all the patents required to manufacture a product. Singer was one of 24 firms using a key set of patents required to manufacture a sewing machine, and the company competed on the basis of marketing networks that included training, servicing and repairs. Singer's monopoly position according to Chandler came from 'the effectiveness of its global organisation. A set of patents without such an organisation could never assure domination; an organisation, even without patents, could' (Chandler 1977: 374). As early as the 1890s, companies began to shift from a reliance on patents for even temporary protection to a process of continuous innovation. This led to the establishment of large in-house research and development departments that meant that 'the research organisations of modern industrial enterprises remained a more powerful force than patent laws in assuring the continued dominance of pioneering mass production firms in concentrated industries' (Chandler 1977: 375).

In his history of American business between 1850 and 1920, Chandler's emphasis on product improvement and innovation suggests research, development and design should become an additional pillar to his three pillars of corporate form – management, marketing and production. He argues that for manufacturers of heavier but relatively standardized machines, 'product improvement and innovation became an even more powerful competitive weapon, far more effective that advertising or canvassing' (Chandler 1977: 419). In effect, competition was not between companies but between the engineers and designers responsible for product development and modification. Even during the 1890s, product innovation was frequently driven by customers, and the sales team would provide engineers with information regarding the specific performance requirements of customers. The engineers would then engage in a discussion with the production department. In this scenario, companies had to ensure that 'the flow of ideas as well as goods had to be co-ordinated' (Chandler 1977: 411). Chandler's account of American capitalism foregrounds the role of the engineer, but neglects the role of the industrial designer. In some respects this reflects the period he was exploring, as industrial design as a named process only really developed during the twentieth century. During the eighteenth century, products were designed by engineers rather than industrial designers. In many cases it is possible to highlight the role played by silent design in the history of industrial capitalism; products are designed by someone, although the individual designer and the design process tend to remain invisible. All products are designed by someone at some stage. I want to argue that there are four pillars of corporate form: management, marketing, design and production. The design pillar functions as a bridge between the marketing and production pillars. The design pillar became critically important as an explicit source of competitive advantage in the twentieth century, but design was an implicit source of competitive advantage in the eighteenth and nineteenth centuries.

During the formative years of corporate capitalism, design, whilst being largely invisible, began to play an increasingly important role in firm-based competition. Politicians and civil servants began to recognize the role design played in national competitiveness, which led to the introduction of design-centred national policy; it is to this that we now turn our attention.

Industrial design and the Industrial Revolution

The Industrial Revolution is associated with the formation of productive capacity that rapidly replaced craft production with production systems based around standardized products that were mass-produced. During the eighteenth and nineteenth centuries, products were developed and modified by a process of silent design. The industrial design processes were undertaken by engineers, inventors and business owners. Many different business models existed that included price-based competition as well as differentiation based on quality of product. The latter included the quality of raw materials and manufacture, the quality of design and the reputation of the producer. The emergence of product design as a discipline distinct from engineering or fine art is a twentieth-century phenomenon closely associated with enhanced competition and the increasing sophistication of consumers. The latter is partly associated with the development of advertising and the mass media; both developments educated consumers and played an important role in forming consumer tastes, preferences and fashions (Bryson et al. 2004). In the nineteenth century, products were designed by engineers and sometimes artists, as the profession of industrial design had yet to emerge. Even during the 1930s it was unusual for a manufacturing firm to employ an internal designer.

Companies tried to develop a competitive advantage in the marketplace, and the business model adopted was influenced and, in many instances, determined by the product markets that a firm was targeting. The competitiveness of firms manufacturing simple utilitarian products was based around the ability of a firm to manufacture articles of quality at prices that would ensure success in home and overseas markets. This strategy was followed by Kenricks, a family business established in West Bromwich (West Midland, England) in 1791, to manufacture hollow-ware (cast-iron pots and pans) and other products. The history of this firm highlights the importance of innovation for the firm, especially innovation in manufacturing processes rather than product design. During the nineteenth century, commercial success was founded upon the ability of Kenricks to improve production methods and distribution, and to pay attention to the activities of competitors (Church 1969: 59). Kenricks had many competitors. There were many foundries, and their equipment and expertise could be applied to the production of products that 'seemed to offer higher than "normal" profits' (Church 1969: 65). Most of the firms in this product-market manufactured pots and pans as well as a great variety of general ironmongery, such as 'hinges, pulleys, coffee mills, smoothing irons, umbrella stands, doors knockers, latches, handles, scrapers, porters, and a large variety of miscellaneous articles included in the trade term "oddwork"' (Church 1969: 65). Competition amongst manufacturers forced firms to develop strategies that would provide them with some form of market differentiation. Differentiation was developed on price and product quality, and also led to the multiplication of minor differences between competing products.

In 1866, according to William Kenrick, the hollow-ware industry consisted of 14 West Midlands firms, a firm located in Glasgow and one in Liverpool (Kenrick 1866: 108). During the so-called 'Great Depression' that occurred after 1872, the British hollow-ware industry began to suffer from foreign competition. Kenricks' competitiveness had been founded upon a patent that had been granted in 1799. This patent was for a process for beautifying hollow-ware by lining cast-iron vessels with vitreous compounds, and this led to the production of enamelled hollow-ware. This patent lasted until 1855 and enabled Kenricks to display enamelled cast-iron hollow-ware at the Great Exhibition at the Crystal Place in 1851 (Church 1969: 63–4). The 'Great Depression' was accompanied by a rise in foreign competition and the realisation that the patterns of many American and French utensils were far superior to those produced in the West Midlands. This difference between the quality of foreign patterns or designs compared to British ones was recognized as early as the International Exhibition of 1862 at which Kenricks displayed their products. This difference highlights the important relationship between competitiveness and product design combined with innovations in manufacturing processes. It is important to remember that during the nineteenth century it was difficult for companies and consumers to acquire information about competing products. The development of colour magazines, television and comparatively inexpensive illustrated books were innovations that were yet to come. A key question is how firms learnt about competitors' products and especially products produced by firms located elsewhere in the UK and also by foreign firms. Another issue involves the question of access to products and sources of design inspiration. In the twenty-first century, it is comparatively simple for individuals and firms to access artefacts and other sources of design inspiration, but this was a difficult task before the development of the mass media and low-cost travel. During the nineteenth century, providing individuals and firms with access to materials that would inspire design improvements in products became one of the central pillars of British design policy.

It is important to distinguish between firms that reacted to competition by making modifications to their products and firms that were proactive in trying to compete on the basis of product design. During the eighteenth and nineteenth centuries, design was a central component in firm-based competitiveness, but for many firms design was a reactive rather than proactive strategy. Government policy developed to try to persuade the majority of firms to realize the strategic importance of product design.

Mathew Boulton, the Soho works and design-based competition

Before exploring the development of design policy during this period, it is useful to explore the activities of a firm that considered design to be a central part of its competitiveness strategy. In 1766 Mathew Boulton founded the Soho Works, a manufacturing plant in Handsworth, to the north of Birmingham, UK. By 1774, Soho produced an astonishing range of products that included 'buckles, spoons, spurs, knife handles, candlesticks, salts, gun furniture, tea-tongs, instrument cases, bottle stands, snuffers, snuffer pans, metal labels, sword hilts, buttons, punch ladles, wine strainers, shoe clasps, whip handles, epergnes, tea vases, coffee pots and lamps, tea pots, canisters, bread baskets, sugar dishes, castors, ice pails, cream jugs, cups, waiters, salvers, table crosses, sauce boats, sacramental plate, tankards, pint and half-pint cups, dishes, plates, tumblers, cheese toasters, skewers, ink stands, toilet plate, fish and putting trowels, bells, etc' (Renshaw 1932: 89). These products were produced by other firms in small workshops in Birmingham; Boulton's innovation was to supervise the manufacture of all these items under the roof of a single factory. In 1775, Boulton entered into a partnership with James Watt, who had made major contributions to the development of the steam engine. The partnership of Boulton and Watt made an extremely important contribution to creating the power that was essential for breaking manufacturing firms' dependence on water as their primary source of power.

Boulton was an entrepreneur who considered that design was an essential factor in the competitiveness of a firm. By the time he was seventeen he had produced a number of innovations in the manufacture of buttons, watch-chains and other small items of metalwork. One innovation concerned the development of inlaid steel buckles which 'were exported in large quantities to France, from whence they were brought back to England and sold as the most recent productions of French ingenuity' (Smiles [1865] 2007: 151). This is an interesting incident that happened around 1745 and highlights an important aspect of industrial production in the West Midlands. Birmingham and the Black Country developed into the workshop of the world (Shill 2006), but during the eighteenth century the region had acquired a reputation for the production of poorly designed and tasteless goods that were known as 'Brummagem' ware. Boulton wanted Soho to produce goods that 'must not only be honest in workmanship, but tasteful in design' (Smiles [1865] 2007: 155). He was aware of the importance of searching out examples of best design and using them to inspire the design of products produced at the Soho Works. Boulton frequently went to London to read and make drawings of objects held in the British Museum, and he began to assemble an art collection. He would also borrow antiques from the aristocracy, and he 'borrowed antique candlesticks, vases, and articles in metal from the Queen' (Smiles, [1865] 2007: 155). He did not confine himself to England, but searched the continent for well-designed objects to copy or to use as the basis for the development of a new design. On one occasion

> he sent his agent, Mr Wendler, on a special mission...to Venice, Rome, and other Italian cities, to purchase for him the best specimens of metal-work, and obtain for him designs of various ornaments – vases, cameos, intaglios, and statuary. On one occasion we find Mr Wendler sending him 456 prints, Boulton acknowledging that they will prove exceedingly useful for the purposes of his manufacture. At the same time, Fothergill [his partner] was travelling through France and Germany with a like object. (Smiles, [1865] 2007: 155–6)

Boulton was trying to improve the quality of the designs produced by his firm. Royalty commissioned articles from him, and encouraged by such royal patronage, Boulton tried to produce extremely high-quality objects. He employed London artists to design some of his best products and also regularly employed French and Italian designers.

The design of Boulton's products was heavily influenced by works of high art. In some cases he produced copies and in others used artefacts and drawings as design inspirations. Firms have always suffered from competitors copying their products and designs, and one could argue that some of Boulton's products were too closely aligned to products produced by others. Companies have always copied from one another, and firms have to try to protect their processes and their designs. During the early years of the Industrial Revolution, intellectual property protection was relatively underdeveloped. Innovations could be patented, but the protection of designs and some processes was problematic. Boulton was aware of this problem, and he played a role in trying to protect Birmingham businesses from unfair copying. In 1783, Boulton played a prominent role in the establishment of the Commercial Committee, the forerunner to the Birmingham Chamber of Commerce. This Committee introduced a reward of £50 Guineas for intelligence that would convict anyone attempting to persuade a Birmingham manufacturer to leave the town to work in a foreign country. The Committee's policy was to attract new capital, expertise and ideas to the city and 'as far as possible to make the door open only from the outside' (Renshaw 1932: 90). New ideas were considered to be equivalent to new capital. In 1786, local manufacturers decided not to admit strangers to their workshops, as they were 'taking as few chances as possible of having their processes copied' (Renshaw 1932: 90). Companies located in Birmingham were sensitive to the loss of commercially valuable information. Thus, when 'Saunders took out the first patent for making covered buttons by dies and pressure he felt compelled to remove from Birmingham to Bromsgrove in order to protect his process from being copied by local competitors' (Renshaw 1932: 90).

The development of manufacturing in Birmingham and the Black Country highlights the importance of design amongst successful firms. Firms competed on process and design, but during the eighteenth and nineteenth centuries design was a silent process. It was silent as the design profession was yet to develop, but it was not 'silent' in the development of national policies that were intended to enhance the ability of British firms to compete on the basis of design rather than price. It is to the development of design-centred economic policy that we now turn our attention.

Incorporating industrial design into national economic policy

On 25 June 1731, 14 Dublin gentlemen met to establish The Dublin Society for Improving Husbandry, Manufactures, and other Useful

Arts, better known since 1821 as the Royal Dublin Society (RDS). The RDS was responsible for founding many of Ireland's most important cultural institutions: the National Library, National Museum, National Gallery and National Botanic Gardens (Craig 1980). It also established the first major school for the arts in the country; this school became the Metropolitan School of Art and subsequently the National College of Art and Design. The RDS investigated, experimented with and encouraged improvements that would enhance the quality of goods manufactured in Ireland. Similar societies were established in Edinburgh and London.

In 1754, the Society for the Encouragement of Arts, Manufactures, and Commence (SEAMC) was founded in London to improve the liberal arts and sciences and to stimulate enterprise by the award of cash prices and metals. In 1761, the SEAMC held its first industrial exhibition of models and machines and began to create a permanent collection of models of inventions and agricultural machinery. Prince Albert (1819–61), the husband and consort of Queen Victoria (1819–1901), became interested in the SEAMC as he considered that it had the potential to operate at the interface between the Royal Society, with its concern with the sciences, and the Royal Academy's engagement with the arts. Although the SEAMC was almost moribund when Prince Albert became aware of it in May 1840, the society was transformed when he became its President in 1843 (Ames 1968: 59–60).

In 1847, the SEAMC was granted a Royal Charter. The Charter was approved by Prince Albert who was then President of the SEAMC. In 1908 the SEAMC was transformed into the Royal Society for the Encouragement of Arts, Manufactures and Commerce (RSA) (Gloag 1946: 96). In 1924, the RSA launched its first industrial design competition to try to introduce manufacturers to the work of young designers. The date of this competition indicates that the RSA was very much aware of the developing profession of industrial design. From its establishment, the SEAMC tried to enhance the relationship between the art and manufacturing worlds.

Perhaps one of the RSA's most important innovations was the proposal in January 1848 to establish a series of national industrial exhibitions. This proposal had been made by Henry Cole (1808–82) to Prince Albert (Ames 1968: 82), and the Prince had informally put the proposition to members of the British government, but no one was interested (Auerbach 1999: 20). According to Ames, 'more preparation of the ground was needed; and both men were very good at such preparation – we might call it propaganda, public relations, or salesmanship' (Ames

1968: 82–3). The result of this propaganda was The Great Exhibition of 1851 – Britain's first international exhibition, created as a celebration of international design and modern technology. The RSA initiated The Great Exhibition and was responsible for raising funds for prizes.

The establishment of SEAMC represents the concerns of a group of private citizens in improving the relationship between the arts and manufacturing. The government took some time to become concerned with the relationship between design and competitiveness, and initially this interest was prompted by an attempt to introduce copyright protection for printed designs. Thus, in 1787, the British Parliament approved a two-month copyright on printed design, and in 1794, extended this protection to three months (Kriegel 2004: 240).

It has always been the case that companies and nations learn from each other and try to maintain and develop competitive advantage. Around 1800, it had been assumed that improvements in the fine arts automatically led to improvements in other areas of design. The rapid development of manufacturing had shown that this assumption no longer held true (Rosenthal 1995: 129). Private patronage had always been important for painters and other artists; however, in the 1830s the British government began to become involved in the relationship between art and manufacture. There were difficulties initially that came from the government's involvement in design. In 1827, George IV had begun a series of improvements to Windsor Castle that included furniture designed by the architect A.C. Pugin. During this period it was unusual for furniture companies to employ designers, but Morel and Seddon (Great Marlborough Street, London), one of the firms commissioned to provide furniture for Windsor Castle, employed four in-house designers and design assistants. A major difficulty ensued in that a Parliamentary committee examined the bills for the work at Windsor and reported to the House of Commons in 1831 that a manufacturer should be his own designer and in consequence would not authorize payment for drawing or designing (Kirkham 1995: 284). This statement echoed the practice of the time that did not distinguish between the craft and design processes. It was assumed that the design for a piece of furniture would originate in the mind of the craft-worker and that design would be included in the final price. The conflation of design with craft did not hold true when an architect designed furniture for a client; in such cases the architect obtained a separate fee as it was recognized that they were design professionals and would not be involved in making the object. The difficulties associated with the settlement of the Windsor Castle expense claim meant that many furniture makers were reluctant to employ designers as they could no longer rely on being able to charge for their work (Kirkham 1995: 284).

The findings of the 1831 Parliamentary committee highlight a failure in some British industries to develop a division of labour between the design function and the process of making or manufacturing a product. According to Kirkham, the separation between craft, business and design developed over the period 1785–1851 in response to the necessity to sell goods at home and abroad (Kirkham 1995: 274). She argues that during this period, design was recognized as a separate profession. This confuses the development of awareness of the importance of design with the development of a design profession. The former developed during this period whilst the latter is more closely associated with the 1920s and 1930s rather than the 1830s (Heskett 1980: 105).

During the nineteenth century British manufacturers tended to concentrate on improving the efficiency of their production processes rather than on product design. The result was that British goods tended to be inexpensive but often unattractive (Auerbach 1999: 11). Many British furniture, textile and pottery manufacturers began to assume that consumers did not want to purchase products designed by British designers and to overcome this problem began 'to spend large sums on foreign designers' (Auerbach 1999: 11). Britain's concern with industrial design can be traced back to the 1830s when a debate developed regarding the poor quality of British design. The 1832 *Reform Act* partially transformed the British Parliament by removing some of the distortions that had previously existed in the election process. This enabled some radical Members of Parliament to be elected who began to raise concerns with the quality of British design. Some British products, especially textiles, were being out-competed by better-designed goods, and in particular, goods designed in France. In a speech to the House of Commons in 1832, Sir Robert Peel stated that

> It is well known that our manufacturers were, in all matters connected with machinery, superior to all their foreign competitors: but in pictorial designs, which were so important in recommending the production of industry to the taste of the consumer, they were unfortunately, not equally successful: and hence they had found themselves unequal to cope with their rivals. (Peel quoted in Read 1934: 13)

The argument was that improvements in the design of British projects would improve sales and profitability.

A Select Committee on Arts and Manufactures was established by Parliament in 1835 to explore the best means of extending knowledge of the art and principles of design amongst the manufacturing population. This report identified a lack of trained designers and a tendency for British firms to import foreign designs, and in the process elevated industrial design into a national preoccupation. The Select Committee collected evidence from manufacturers, designers and artisans. It identified that 80 Schools of Design had been established in France and 33 just in Bavaria. Both Germany and France were identified as countries that had developed enlightened policies to improve industrial design. Britain decided to copy these policies, and as a consequence, the Committee's Report of 1836 recommended that government schools of design should be established throughout the UK and that more art galleries should be opened and exhibitions held to educate consumers (Kirkham 1995: 285). A government committee was established with the power to awards grants for the establishment of public schools of design. In Birmingham, the Society of Arts made a successful application and opened a school in October 1843. This school continued until 1851 when quarrels amongst the teaching staff led to the retirement of the head. The Board of Trade recommended the appointment of George Wallis, who had worked for the Committee of the Great Exhibition of 1851 (Waterhouse 1954: 15). In their early years, the schools for design provided a teacher training programme for young boys who, when qualified, intended to teach in elementary schools. The schools taught drawing and tended to neglect the artistic or creative side of the design process. The design schools failed to break down the barriers between 'art' and 'industry' and this led to artists with no experience of commerce or manufacturing processes being employed. The schools failed to provide the training required to educate designers to appreciate the relationship between materials, production methods and 'art'.

The Select Committee also encouraged Parliament to revisit the copyright issue. In 1839, stronger copyright legislation was approved that established a central designs registry in London (Kriegel 2004: 240). In 1849 a Select Committee of Parliament was appointed to explore the effectiveness of the Schools of Design. The committee found that 16,000 students had been enrolled in the schools over a 13 year period, but it was unclear if the schools had done anything to improve the design of manufactured products.

Since 1835, Britain has periodically attempted to develop design-related national and regional policy. The 1851 Great Exhibition was intended to improve the overall quality of the design of British products. In 1931, the Board of Trade appointed another Committee to explore the relationship between art and design. This report suggested that the best artists and craftspeople should be encouraged to turn their energies towards industrial manufacturing. The 1943 unpublished *Weir Report on Industrial Design and ART in Industry* recommended that a Council for Industrial Design (COID) should be established to stimulate improvements in design in industry that would enhance exports. The COID organized the Britain Can Make It (BCMI) Exhibition in 1946 in which 5000 objects were displayed that were chosen to try to convince manufacturers, retailers, consumers and foreign buyers that design was of critical importance in the post-war period.

All these British policy interventions that were intended to stimulate the design-intensity of creative and industrial design returned to the British political agenda in the March 2005 budget statement in which the Chancellor of the Executor, Gordon Brown, (HM Treasury 2005: 68) commissioned research that would consider how to raise awareness of the potential for SMEs (Small and Medium-sized enterprises) to draw on the UK's world-leading creative skills in areas such as industrial design.

Industrial design policy and the establishment of design exhibitions

Industrial design has been incorporated into economic policy through educational programmes, targeted subsidies to encourage firms to enhance the design-intensity of their products and exhibitions intended to educate consumers and producers. The history of design exhibitions is complex and can be traced back to France towards the end of the eighteenth century. In 1798, the Marquis d'Aveze assembled a collection of objects which he displayed throughout the house and gardens of the Maison d'Orsay, Rue de Varennes. This bazaar was a commercial venture intended to sell the objects on display. The success of this venture encouraged the French government to erect a Temple of Industry on the Champ de Mars that would be used to display the most beautiful object manufactured in France. A system of juries was established to judge the comparative merits of objects being considered for inclusion in the exhibition. The success of this first Exposition encouraged the French government to repeat it annually, but the events of this period meant that the second exposition was held in 1801 (Art-Journal 1970: xii). The tenth Industrial Exposition was held in 1844 and included objects produced by 3960 manufacturers whilst the eleventh was held in 1849 with 4494 exhibitors (Art-Journal 1970: xii). These early French exhibitions are associated with the Napoleonic Wars and were considered as 'economic weapons in the fight against England' (Auerbach 1999: 9). Most of these exhibitions were directed at foreign investors and consumers, but excluded participation by foreign manufacturers.

In Britain, many small bazaars had been organized for the display and sale of goods produced by manufacturers based in a specific locality. These were not design exhibitions *per se*. Manchester, Leeds and Dublin had held such exhibitions, but the first building erected in Britain solely for the exhibition of manufactured goods was constructed in Birmingham in 1849 for the annual meeting of the British Association (British Association for the Advancement of Science 1849; Art Journal 1970: xii). There had been an earlier, unsuccessful attempt to establish a national exhibition of industrial products. In 1828, an exhibition committee was formed and George IV agreed to be patron. The proposal was to hold annual exhibitions of new and improved products, but the scheme failed due to inadequate support from industry (Gloag 1946: 96).

The RSA played a key role in persuading British manufacturers to participate in industrial exhibitions. In 1844, the Society reinstated its annual exhibition and prize competition to encourage manufacturers to create tasteful objects, but also to raise money for the society. This was unsuccessful, but the Society persevered and organized exhibitions in 1845 and 1846. The Society found it difficult to persuade manufacturers to participate in these events and only 800 people attended the 1845 exhibition. Manufacturers were reluctant to participate in the exhibitions as many failed to appreciate the commercial advantages that could be gained from participation in a major industrial exhibition. There were many reasons why manufacturers refused to participate. First, many firms were small and did not have the resources to participate. Second, manufacturers considered that participation in an exhibition might damage them commercially as competitors would have an opportunity to steal their ideas (Auerbach 1999: 15). Third, transportation throughout Britain was difficult and this may have made the proposition of a national exhibition appear to be unrealistic. Fourth, manufacturers may have considered that they were successful and did not require the stimulus that would come from an exhibition.

In 1846, the Society organized another exhibition. At this exhibition a simple white tea service designed by Henry Cole was exhibited and awarded a prize (Auerbach 1999: 17). Prince Albert purchased this tea set for Buckingham Palace while Cole joined the Society of Arts and was quickly promoted to the council. Cole played an important role in the 1847 exhibition by persuading manufacturers to lend articles for exhibit. Around 20,000 people attended this exhibition and its success encouraged manufacturers to participate in the 1848 exhibition. The 1847 exhibition represented a turning point in the development of British capitalism. It was around this time that manufacturers realized that they had to begin to create a market for their goods and that consumers became to discriminate between competing products. It is important to remember that a mass market for consumer goods did not exist at this stage and that advertising was still in its infancy.

The success of the 1847 and 1848 exhibitions led to the decision to establish a Royal Commission to organize a Great Exhibition that would be held in 1851. This exhibition was an attempt to improve the quality of British design by exhibition, competition and encouragement. There were more than 100,000 exhibits sent in by almost 14,000 individual and corporate exhibitors, and exhibits came from Britain, its colonies and dependencies as well as many other countries (Auerbach 1999: 92). The exhibits were classified into four types: raw materials, machinery, manufactured goods and fine arts. The classification system highlighted the importance of manufacturing, but was also an attempt to educate manufacturers and consumers about the benefits of good design. The primary problem regarding British manufacturing was not the manufacturing process, but the failure of companies to appreciate the importance of decoration, ornamentation and good design.

The Great Exhibition attracted over six million visitors and involved over 15,000 exhibitors. Fifty per cent of the exhibition space was devoted to 7351 British and Empire products and 6556 products from overseas. The Great Exhibition generated a surplus of £186,437. There was much discussion and debate over this surplus, but eventually it was decided to purchase two landed estates (35 hectares) in South Kensington, London, and to construct a building to house the exhibits given to the Royal Commission by exhibitors. The government contributed a grant of £150,000 and also £15,000 towards building costs. The new building was opened in 1857 and was called the South Kensington Museum. In 1909, this museum became the Victoria and Albert Museum. The museum was intended to collect examples of good design and to exhibit them in such a way as to encourage manufacturers and consumers to appreciate their qualities. The South Kensington site was used for the 1862 International Exhibition and for the 1886 Colonial and Indian Exhibition. Ultimately, the surplus from the Great Exhibition created the South Kensington educational and museum complex that includes the Victoria and Albert Museum, the Science Museum, the Natural History Museum, Imperial College of Science and Technology and the Royal Albert Hall.

After the 1851 exhibition, around 31 large-scale temporary international exhibitions were held in the British Isles. Single-industry exhibitions were also held, for example, in Birmingham in 1865 for the metal industry. County exhibitions were held as well. In 1881, a committee was established to organize a Worcestershire Exhibition (Worcestershire Exhibition 1882). This was held in 1882 and consisted of four courts: fine arts and historical, industrial, industrial machinery in motion and a display of carpets and carpet-making machinery. In the industrial machinery in motion court, the visitor to the exhibition could experience the sights, sounds and smells of industrial production. Steam engines had been set up to provide the exhibits with power by overhead transmissions and visitors could observe, for example, 14 machines, engaged in the manufacture of boots and shoes. The whole sequence of leather glove manufacture and carpet manufacture was also on display as was the weaving of horse-hair mats and the manufacture of wire netting. The Worcestershire exhibition lasted three months, and attracted nearly a quarter of a million people and generated a surplus of £1,867 9s 6d. The surplus was used to clear a deficit that had developed in the accounts of the public library as well as to construct the Victoria Institute, a building that contained a public library, a school of art, a museum, a technical school and an art gallery.

The Great Exhibition also encouraged other counties to organize great *expositions universelles* (Table 2.1). British companies participated in many of these events, as the exhibitions were important for developing export markets. Throughout the second half of the nineteenth century, Britain, alongside other nations, built many pavilions at these *expositions universelles* that were designed to reflect national styles. Grand exhibitions became an important mechanism for the construction and display of national identity. They allowed a nation to display the quality, diversity and strength of its manufacturing base as well as to make a series of statements regarding national design, art, engineering and architecture.

The importance of organized exhibitions did not dissipate in the twentieth century. In 1924, the British Empire Exhibition, Wembley, London, attracted over 17 million visitors. The emphasis in this exhibition was on engineering and technical skills rather than on the quality of the design. In 1946, the Britain Can Make It Exhibition (BCMI) was held at the Victoria and Albert Museum. This exhibition was organized by the COID, and was intended to enhance Britain's export trade and at the same time encourage manufacturers and consumers to focus on the benefits that come from well-designed products. This was an extremely selective exhibition with the COID-appointed selection committee rejecting over

Table 2.1 Selected International Exhibitions held during the reign of Queen Victoria (1837–1901)

Date	Location	Exhibition Name
1851	London	Great Exhibition of the Works of Industry of All Nations
1853	New York	Exhibition of the Industry of All Nations
1853	Paris	Exposition Universelle
1853	Dublin	The Great Industrial Exhibition
1862	London	International Exhibition
1865	Dunedin	New Zealand Exposition
1865	Dublin	International Exhibition of Arts and Manufacture
1866	Melbourne	Intercolonial Exhibition
1867	Paris	Exposition Universelle
1871	London	International Exhibition of Arts and Industries
1872	London	International Exhibition of Arts and Industries
1873	Vienna	Weltaustellung
1873	London	International Exhibition of Arts and Industries
1874	London	International Exhibition of Arts and Industries
1876	Philadelphia	Centennial International Exposition
1878	Paris	Exposition Universelle
1879–80	Sydney	International Exhibition
1880–1	Melbourne	International Exhibition
1883	Amsterdam	International and Colonial Exhibition
1883–4	Boston	American Exhibition of the Products and Manufacturers of Foreign Nations
1883–4	Calcutta	International Exhibition
1884–5	New Orleans	The World's Industrial and Cotton Centennial Exposition
1885	Antwerp	Exposition Universelle
1886	Edinburgh	International Exhibition of Industry, Science and Art
1887–8	Adelaide	Jubilee International Exhibition
1888	Barcelona	Exposició Universal
1888	Glasgow	International Exhibition
1888–9	Melbourne	Centennial International Exhibition
1889	Paris	Exposition Universelle
1889–90	Dunedin	New Zealand and South Seas Exhibition
1891	Kingston	Jamaica International Exhibition
1883	Chicago	World's Columbian Exhibition
1894	Antwerp	Exposition International
1894	San Francisco	California Midwinter International Exposition
1894–5	Hobart	Tasmanian International Exhibition
1897	Brussels	Exposition Internationale
1897	Guatemala	Exposition Internationale
1897	Nashville	Tennessee Centennial and International Exposition

Continued

Table 2.1 Continued

Date	Location	Exhibition Name
1897	Stockholm	All Männa Konst-och Industriutställingen
1900	Paris	Exposition Universelle
1901	Buffalo	Pan-American Exposition
1901	Glasgow	International Exhibition
1901	Calcutta	Indian Industrial and Agricultural Exhibition

two-thirds of the 15,836 goods that were submitted by industry for consideration for inclusion in the exhibition (Woodham 1997).

The COID was established in 1944 as a state-funded body to promote good design amongst British manufacturers. It was an attempt by Britain to recover market share that Britain had lost to American companies during the Second World War. During the 1930s, many American companies had discovered and experimented with the relationship between product design and competitiveness. The COID was tasked with promoting good design in British industry by disseminating design advice to producers and to government departments. It was also tasked with running design exhibitions, with providing advice on design training and education and with educating consumers to identify and appreciate good design. This was an important policy development as the COID was the first state-sponsored body tasked with the development of a design-centred policy toolkit. It was influential as many other countries copied the COID and established their own state-funded design promotion organisation. In 1967, the remit of COID was broadened to include engineering and capital goods. A new Council for Engineering Design was established but in 1972 it was decided to create the Design Council, a new single body that would be responsible for engineering and design.

The BCMI exhibition was dubbed in the popular press as the 'Britain Can't Have It Exhibition', as many of the goods on display were prototypes or were available for export only given the war-time restrictions on raw materials that were still in force (Heskett 1995: 294). In 1951, another major exhibition was organized in which the COID played an important organisational role. This was the Festival of Britain and it was originally conceived as an international exhibition to celebrate the centenary of the Great Exhibition. Nevertheless, the uncertain economic climate led to the proposed international exhibition's conversion into a national celebration of Britain's past, present and future contributions to societal, cultural and technological progress.

Conclusion

The relationship between design and firm-based competitiveness has always been an important feature of capitalism. Design, however, has often been silent or has been something that as been added to products as an afterthought. During the nineteenth century, many British firms clothed 'manufactured articles with "applied art" [and this] debauched the whole conception of design' (Gloag 1946: 94). Design must be an integral part of the production process of products and services.

It is important to appreciate that the process of industrial design was formalized in the twentieth century with the emergence of professional designers. Prior to the 1920s, the focus of many policy debates was on the relationship between manufacturing and art rather than between manufacturing and industrial design. What can be identified is a shift from technological- to design-centred competitiveness. During the nineteenth century, many companies acquired a competitive advantage on the basis of the efficiency of their production processes and their ability to compete on price and quality rather than on price, quality and design. During the early nineteenth century, many markets were highly localized and competition from foreign manufacturers was restricted. Competition from foreign producers began to become important and companies had to begin to consider other ways in which to differentiate themselves in the market place. This is a complex issue. Firms were able to develop brands, distinctive products that were protected by patents and to a lesser extent design registrations and trade marks. Companies producing products that had little or no legal protection had to compete on price and/or on design. This attempt by firms to acquire market differentiation is reflected in Chandler's three pillars of corporate form – management, marketing and production. In this chapter I have argued that, in fact, there are not three pillars of corporate form but four: management, design, marketing and production. Firms can compete by investing in each of these activities and some firms place more emphasis on one or more of these pillars.

The relationship between industrial design and national competitiveness is complex and also extremely important. It is complex as the relationship is founded upon the historical accumulation of national design associations. By this it is meant that a nation's design identity is constructed over a long time period and is produced by the actions of many different firms with perhaps some public policy input. In part, a nation's design identity is produced by consumer experience and by the stories that are told by consumers about their experience of particular products (Rusten et al. 2007). These design stories are in part created by design critics and the design media industry. Education must be a central feature of all design-informed national policies. The question is education for whom? Firms must be educated to appreciate the importance of industrial design for firm-based competitiveness and consumers must be educated to appreciate well-designed products. It is important to appreciate that this distinction between the education of the consumer and the producer is a very old one. In 1934, Read argued that:

> the production of well-designed objects of daily use would undoubtedly be stimulated by a higher level of public taste, by an increased sensibility to the elements of design. This is one side of the problem, and certainly an important one: we will call it *the education of the consumer in aesthetic appreciation*. The other side of the problem is more obvious: we will call it *the education of the producer in aesthetic design*. (Read 1934: 163)

Both types of education are equally important. Well-informed consumers will force or encourage manufacturers to produce well-designed products and such products must be commercially viable. The establishment of design exhibitions to showcase best practice in design and the creation of museums were old solutions to this educational issue. New solutions are embedded in the primary and secondary school system and in the education of designers. The design of products is complicated by geography and much more research is required to understand the ways in which companies design products to meet the needs of particular consumers. Geography creates design complexity; many products need to be designed to meet the needs of particular markets and this requires an appreciation of the differences that exist between national consumer cultures. The implication is that national design policies must be targeted at encouraging firms to enhance the design-intensity of their products. Design policy must also operate to ensure that firms modify their designs to meet the requirements of specific markets. In this case, the interaction that occurs between geography and design is a vital element in the competitiveness of firms and nations.

References

Ames, W. 1968. *Prince Albert and Victorian Taste*, Chapman & Hall, London.
Art-Journal 1970. *The Crystal Palace Exhibition: Illustrated Catalogue London 1851 – An Unabridged Republication of the Art-Journal Special Issue*, Dover Publications, New York.

Auerbach, J.A. 1999. *The Great Exhibition of 1851: A Nation on Display*, Yale University Press, New Haven.

Braverman, K. 1974. *Labor and Monopoly Capital: The Degradation of Work in the Twentieth Century*, Monthly Review Press, New York.

British Association for the Advancement of Science 1849. *Catalogue of the Articles in the Exhibition of Manufactures and Art, in Connection with the Meeting of the British Association for the Advancement of Science at Birmingham*, M. Billing's Steam-press, Birmingham.

Bryson, J.R. 2000. 'Spreading the message: management consultants and the shaping of economic geographies in time and space'. In Bryson, J.R., Daniels, P.W., Henry, N. and Pollard, J. (eds) *Knowledge, Space, Economy*, 157–175, Routledge, London.

Bryson, J.R., Daniels, P.W. and Warf, B. 2004. *Service Worlds: People, Organisations, Technologies*. Routledge, London.

Bryson, J.R. and Rusten, G. 2008. 'Transnational corporations and spatial divisions of "Service" expertise as a competitive strategy: The example of 3M and boeing', *The Service Industries Journal*, 28:3, 307–323.

Bryson, J.R., Taylor, M. and Daniels, P.W. 2008a. 'Commercializing "Creative" expertise: business and professional services and regional economic development in the West Midlands, UK', *Politics and Policy* 36:2, 306–328.

Bryson, J.R., Taylor, M. and Cooper, R. 2008b. 'Competing by design, specialization and customization: manufacturing locks in the West Midlands (UK)', *Geografiska Annaler: Series B, Human Geography* 90:2, 173–186.

Chandler, A.D. 1977. *The Visible Hand: Managerial Revolution in American Business*, The Belknap Press of Harvard University Press, Cambridge, MA.

Church, R.A. 1969. *Kenricks in Hardware: A Family Business: 1791–1966*, David & Charles, Newton Abbot.

Craig, M. 1980. *Dublin 1660–1860*, Allen Figgis, Dublin.

Dicken, P. 2003. *Global Shift: Reshaping the Global Economic Map in the 21st Century*, Sage, London.

Finer, A. and Savage, G. (eds) 1965. *The Selected Letters of Josiah Wedgwood*, Cory, Adams and MacKay, London.

Forty, A. 1995. *Objects of Desire: Design and Society Since 1750*, Thames and Hudson, London.

Gloag, J. 1946. *Industrial Art Explained*, George Allen and Unwin, London.

Gloag, J. 1949. *Time, Taste and Furniture*, The Richards Press, London.

Heskett, J. 1980. *Industrial Design*, Thames and Hudson, London.

Heskett, J. 1995. 'Industrial design'. In Ford, B. (ed.) *Modern Britain*, 288–318, Cambridge University Press, Cambridge.

HM Treasury-2005. *Investing for our future: Fairness and opportunity for Britain's hard-working families*, HM Treasury, London (Budget Report, March).

Kenrick W. 1866. 'Cast-iron hollow-ware, tinned and enamelled, and cast-ironmongery'. In Timmins, S. (ed.) *The Resources, Products and Industrial History of Birmingham and the Midland Hardware District*, 103–109, Robert Hardwicke, London.

Kirkham, P. 1995. 'The "Applied arts": design, craft and trade'. In Ford, B. (ed.) *The Romantic Age in Britain*, 72–290, Cambridge University Press, Cambridge.

Kriegel, L. 2004. 'Culture and the copy: calico, capitalism, and design copyright in early victorian britain', *Journal of British Studies* 43, 233–265.

Lucie-Smith, E. 1981. *The Story of Craft: The Craftsman's Role in Society*, Phaidon, Oxford.
NESTA 2009. *Attacking the Recession – Setting the Agenda for a More Innovative Economy: How a More Innovative Economy Can Beat the Recession: A Three-Point Plan for the UK*, NESTA, London.
Petty, W. [1678], 2004. *Essays on Mankind and Political Arithmetic*, Kessinger Publishing: Whitefish, MT.
Read, H. 1934. *Art and Industry: The Principles of Industrial Design*, Faber and Faber, London.
Renshaw, T.L. 1932. *Birmingham: Its Rise and Progress*, Cornish Brothers, Birmingham.
Rosenthal, M. 1995. 'The fine arts'. In Ford, B. (ed.) *The Romantic Age in Britain*, 118–171, Cambridge University Press, Cambridge.
Rusten, G., Bryson, J.R. and Aarflot, U. 2007. 'Places through product and products through places: Industrial design and spatial symbols as sources of competitiveness', *Norwegian Journal of Geography* 61:3, 133–144.
Shill, R. 2006. *Workshop of the World: Birmingham's Industrial Legacy*, Sutton Publishing, Gloucestershire.
Smiles, S. [1865] 2007. *Lives of Boulton & Watt*, Nonsuch Publishing, Gloucestershire.
Smith, A. 1977. *The Wealth of Nations*, Skinner, A. (ed.), Pelican, Middlesex.
Taylor, M. and Bryson J.R. 2006. 'Guns, firms and contracts: The evolution of gun-making in birmingham'. In Taylor, M. and Oinas, P. (eds) *Understanding the Firm: Spatial and Organizational Dimensions*, 61–84, Oxford University Press, Oxford.
Thompson, D. 2008. *Watches*, The British Museum Press, London.
Waterhouse, R.E. 1954. *The Birmingham and Midland Institute: 1854–1954*, The Birmingham and Midland Institute, Birmingham.
Woodham, J.M. 1997. 'Britain can make it and the history of design'. In Maguire, P.J. and Woodham, J.M. (eds) *Design and Cultural Politics in Postwar Britain*, 17–28, Leicester University Press, Leicester.
Worcestershire Exhibition 1882. *Official Catalogue*, W.E.Tucker & Co, Worcester.
Wornum, R.N. [1851] 1970. 'An essay on ornamental art as displayed in the industrial exhibition in Hyde Park, in which the different styles are compared with a view to the improvement of taste in home manufacturers', in Art-Journal, *The Crystal Palace Exhibition: Illustrated Catalogue London 1851 – An Unabridged Republication of The Art-Journal Special Issue*, i-xii, Dover Publications, New York.

3
Better by Design? A Critical Appraisal of the Creative Economy in Finland

Nikodemus Solitander

> Design is a creative activity whose aim is to establish the multi-faceted qualities of objects, processes, services and their systems in whole life cycles. Therefore, design is the central factor of innovative humanisation of technologies and the crucial factor of cultural and economic exchange.
>
> (The International Council of Societies of Industrial Design ICSID, http://www.icsid.org, accessed October 2008)

Introduction

In Finland the design sector evokes a mythical and romanticized image of the small nation overcoming the hardships of World War II through the establishment of a national identity developed around design and architecture, and successfully transforming this identity into an export-commodity, in the form of cultural products. Historically the creation of 'Finnish Design' was also in many respects similar to that of 'Scandinavian Design', a conscious effort of branding, promoted through international touring exhibitions, expositions and competitions (Stenros 2004). 'On an international scale Finnish design products struck a chord with consumers in the 1950s and 1960s as they mirrored the spirit of the times and the changes in Western society': for example, women's equality, urban migration, and modern lifestyles (Stenros 2004). In many respects, however, the impact of Finnish design was perhaps more important on an imagined level and in the construction of subjects – rather than having a visible impact on exports. In the late 1950s and 1960s traditional heavy industries such as forestry dominated exports almost completely (Oinas 2005). In modern times, especially throughout the 1980s and much of the 1990s, the design sector has in many ways struggled to live up to the legacy of the international success that was acquired in the 1950s and 1960s both in terms of international recognition and in forming Finnish identity. Design remained something of a myth in the Finnish economic landscape, present in the background but not at the centre of economic discourse.

After the turn of the last century, however, the design sector was propelled to the centre of Finnish economic policy initiatives, both on the regional and national level and in imagination but also in practice. As I will argue in this chapter, the re-emergence of design into the economic landscape is inherently tied to the rise of a new spirit of creative capitalism in which 'creative destruction, chic slacker-cool, and designer resistance are now celebrated in organization[s] advocating freedom around normative inputs' (Fleming and Spicer 2008: 303). Following the much-cited emergence of a 'network society' and a 'knowledge society' (e.g. Drucker 1993; Castells 1996) creativity is heralded as the next big thing that will transform the economic landscape of Western regions. The establishment of a 'creative economy' has become an imperative for nations seeking to be competitive in the global economy (Thrift 2005; Jeanes 2006; Jones and Boon 2007). The creativity gospel is eagerly reproduced and diffused by various academic disciplines, ranging from urban studies (e.g. Florida 2002; Landry 2000) to economic geography (e.g. Desrochers 2002; Drake; Scott 2004; Power and Hallencreutz 2004; Törnqvist 2004) and to management and organization studies (e.g. Gibb and Waight 2005; Wilson and Stokes 2005; Bilton 2006)[1]. At the centre stage of the diffusion of creative capitalism stands the writings of Richard Florida (Gibson and Klocker 2004; Peck 2005). Florida's texts (e.g. 2002, 2005, 2006b) relay the essence of the creativity discourse: the (Western) world is in the midst of a revolutionary economic transformation, societies are moving away from the uninspiring practices of a service economy towards a cool and caring creative economy, one that generates wealth by harnessing intellectual labour, intangible goods and human creative capabilities. This mantra has found a willing audience as policymakers are increasingly looking towards creativity for regeneration, employment opportunities, and most importantly, competitive advantages (Porter 1990). A recent report commissioned by the European Commission states that 'there is a competitive race to attract talent and creators ("the creative class") to localized environments supporting the clustering of creativity and innovation skills' (KEA European Affairs 2006). The regional 'race for talent' has produced many strategy projects that attempt to address the link between place, competitiveness and creativity. On the city-regional level there are projects such as 'Creative London', (http://www.creativelondon.org.uk), 'Creative Tacoma' (http://www.creativetacoma.org) and the 'Creative Industries Precinct' in Queensland, Australia; on the national level there is the 'Better by Design' programme in New Zealand (http://www.betterbydesign.org.nz/), Singapore has initiated a 'Creative Industries Development Strategy' (http://www.mica.gov.sg/mica_business/b_creative.html) and UNESCO has established an international network called The Creative Cities Network (http://www.unesco.org). In this new creative age, industrial design and designers are elevated to the centre stage of discourse. Design and especially industrial design often serve as a link between culture and economy, art and industry, and between 'soft' and 'hard' capitalism. Industrial design combines discourses and language of management, technology and arts. This has also enabled an appropriation of the language (and critique) of art to further strengthen neoliberal discourse, something that I will further explore and elaborate on in this chapter.

The creativity discourse is increasingly, and even particularly, evident in Finland where policy makers have been eager to refer to best-in-creativity-class rankings presented by international research (e.g. Florida and Tinagli 2004; Sorvali et al. 2006; World Economic Forum 2005). In 2004, Finland took steps towards producing a national creativity strategy in order to 'develop the prerequisites of creative activity' (Ministry of Education 2004: 5), the strategy was described as a 'key strategy in international competitiveness' (Koivunen 2004: 5)[2]. In 2002 Finland also launched a national design programme, which aimed to improve the competitiveness and internationalization of the design sector. Design is also an explicit part of the Finnish national system of innovation (Järvinen and Koski 2006). Industrial design has historically been at the heart of what Finns have considered being a national competitive advantage, but the infusion of creativity as a central sign in the language of business and management has further cemented and revitalized its position in the Finnish economic landscape. While it is easy to be sympathetic to the promises of a creative economy and the rise and centrality of industrial design in making the economy more human-centred, more beautiful and safe (e.g. Järvinen and Koskinen 2001: 10), there are pervasive reasons for research to analyze critically the celebratory discourses they are (re)producing, the emerging tensions and the unexplored silences. As ten Bos (2006: 30) reminds us, there is 'some strangeness in a world that generally does not manifest any doubt whatsoever about the goodness of its own endeavour.' In many ways the creative economy and by extension industrial design are part of such a world.

In this chapter, I critically examine, mainly by drawing on insights from poststructuralist critique (e.g. Thrift 2005; Dean 1999; Rose 1996), the discourses, practices and impacts of the creative economy, and specifically in terms of how design and industrial design are presented and used in discourse. By doing this, I wish to emphasize the necessity of analysing the rise of industrial design within the larger discourse of the creative economy. With creativity discourse I largely refer to texts that position creativity as a production factor in relation to (international) competitiveness. I am particularly interested in the role of space in constructing a new subject position (the creative class) based on 'a new creative ethos'; this is illustrated in this chapter through the re-emergence of the design industry in the economic landscape in Finland. The chapter is thus situated in debates on-going in the literature of economic geography. The aim is to expose the complex and competing forces shaping creativity discourses, and thus help researchers and practitioners to ask important questions about the representation of spaces of creativity, and how these are used to produce collective bodies and identifications, foremost of which is the notion of a 'creative class' (e.g. Florida 2002). The chapter is structured as follows: in the first part, I critically analyze the re-emergence of the Finnish design economy in the context of creativity discourse and the subject positions it produces, with a specific emphasis on the naturalization of creativity discourse on various territorial scales in Finland. In the second part, I illustrate the practice of creativity discourse by analysing the establishment of the 'City of Art and Design' in Helsinki, Finland, putting special emphasis on the role of space in the creation of creative subjects.

The rise of creativity discourse

By combining the language of technology, business-thinking and arts and culture, industrial design becomes a central signifier and illustration of what is referred to as the creative economy. A recent report by UNCTAD (2008: iii) on the creative economy states that

> ... the interface among creativity, culture, economics and technology, as expressed in the ability to create and circulate intellectual capital, has the potential to generate income, jobs and export earnings

while at the same time promoting social inclusion, cultural diversity and human development. This is what the emerging creative economy has already begun to do as a leading component of economic growth, employment, trade, innovation and social cohesion in most advanced economies.

Such language also mirrors the essence of industrial design, as design is posited to provide a 'competitive advantage to firms...creating additional value for customers...as well as "humanifying" technology' (translated from Lindström et al. 2006: 29) and it also helps to make the 'world a better, more beautiful, a safer, and a smoother place to live in' (Järvinen and Koskinen 2001: 10). Thus,

> The Finnish world known designers and architects Alvar Aalto, Tapio Wirkkala and Kaj Franck were all humanists. Their work was based on the respect of nature and draw its inspirations from the best traditions and achievements of human culture.... Design for all, or putting people's needs first has become [an] important mission of the European Union. It sees that it is both socially necessary and economically beneficial to build an accessible and barrier free society. It increases equality and expands market possibilities. This mission is shared by the Finnish technology giant Nokia, which sees technology as an enabler of sustainable development. (Sotamaa 2006b: 6)

Industrial design is thus envisioned to both support the essence of capitalism's unlimited economic growth – but also provide it with a soft makeover, picking up the pieces where capitalism has failed as an enabler of 'sustainable yet unlimited' economic growth. Art provides the 'humanist face' to capitalism and technology acts as the enabler. But in the new economy the humanists 'respecting nature and advancing culture' are not individuals such as Aalto and Wirkkala but multinational corporations such as Nokia. In this discourse free society and democracy is equalled with 'expanded market possibilities'. Industrial design acts as an important mediator between seemingly competing discourses, such as economy and art. I will elaborate on this point further, but will begin by presenting the historical background to the rise of the creative economy, before further showing how this also contextualizes the (re)emerging interest in industrial design.

Central to globalization discourses is a focus on increased and accelerated circulation of people, corporations, capital, commodities and images (Brenner 1999). Geographical places, hopelessly immobile and unable to circulate, are mooted to the outskirts of the language of modernity (Zukin 1991). When information and transport technologies increased mobility and decreased geographical transaction costs, for some it also implicated a discursive erasure of place (Cairncross 1997; O'Brien 1992). Economic geography fiercely contested arguments that the end of geography had occurred; in such contestations place was infused as a central sign in globalization discourses, and the rearticulation invested new meanings to geographical places. In the knowledge economy, the mobile actors, most visibly multinational corporations, gravitate towards places that create and animate the 'sources of competitive advantage' (Porter 1990).

In his influential book *The Competitive Advantage of Nations*, Michael Porter suggests that shifts in geographical scales 'represent a new way of thinking about national, state, and local economies' (Porter 2000: 16). From this perspective followed an understanding that globalization is not a discursive eradication of geography, but rather a shake up of the territorialization process. The geographical scales we observe are 'no longer spatially co-extensive with the nationally organized matrices of state territoriality' (Brenner 1999: 435), which had for so long been the naturalized understanding of capitalism's economic geographies. In the knowledge economy, the headquarters of multinational corporations, their research and design departments, and production units, are spreading across national borders with increased ease, but simultaneously they are agglomerating in particular regions (Dicken 2007). This idea is reproduced in various geographical concepts such as innovative milieu (Camagni 1991), industrial districts (Piore and Sabel 1984), clusters (Porter 1990, 2000), and world/global cities (Friedmann 1986; Sassen 1990). Each of these concepts is constituted in relation to (international) *competitiveness*. The all-pervasive, yet abstract quality of competitiveness has become a hegemonic category referring to 'the capacity of a firm to compete, grow and be profitable in the marketplace' (Bristow 2005: 287). When applied to regions and more recently to city-regions competitiveness becomes even more abstract. Yet with its increased hegemony, and naturalized acceptance and understanding, it becomes almost impossible to question (Schoenberger 1998). A strategy report on the Helsinki Metropolitan Area states that the evolution of cities '...depends on their *international competitiveness*. Earlier, urban areas were *"subordinate"* to *national policies* of growth and competition.... Finnish urban regions, too ... are specializing in internationally competitive expertise to create a unique and sustainable role in the *international scramble between urban region's*.' [emphasis added] (Kulkki 2004: 19).

By this understanding competition is increasingly occurring between urban regions, which calls for 'spatial selectivity' (Brenner 2004); political strategies that target particular regions in order to position them 'optimally' in the global 'scramble' for mobile capital.

A recurring critique against the reformulation between competitiveness and regions, most notably the form popularized by Porter (1990), has addressed the narrow focus on economic imperatives compared with concerns over standards of living, welfare and redistribution (Bristow 2005; Schoenberger 1998; Bunnell and Coe 2005). But this critique is increasingly appropriated through the fusion of the economic, social and cultural landscape into the language of capitalism. This can be seen as what Thrift (2005: 4) refers to as 'performative capitalism', wherein a defining feature is capitalism's ability to be highly adaptive and a constantly mutating formation. It includes adopting the language of its critique (Thrift 2005). Industrial design is a case of point: while having strong technological and commercial aspects it also has strong associations to the arts, individuals and the cultural sphere in general. In creative economy discourse the language of economic success is presented as emanating from 'soft' characteristics, such as knowledge, learning and creativity, rather than straightforward technological or cost advantages (Heelas 2002; Ray and Sayer 1999; Thrift 2005). Such language is becoming increasingly dominant in both policy making and academia. A strategy document prepared as a basis for a creative strategy for the city of Toronto declares that the:

> ...development of the creative economy...enhances a city's quality of place, helps to reclaim and revitalize neighbourhoods, enables more innovative thinking and problem-solving across all sectors of the economy.... Creative and cultural activity is also a powerful vehicle for community development and engagement, providing opportunities for economically disadvantaged neighbourhoods and social groups ('Imagine a Toronto...Strategies for a Creative City' 2006: 7).

The creative economy is described as subversive, a vehicle of empowerment for the 'economically disadvantaged', a way to 'reclaim neighbourhoods', all while reproducing a neoliberal syllabus. In line with the images of dressed-down, caring and compassionate capitalism, design is also often portrayed through its will to make 'the world a better, more beautiful, a safer, and a smoother place to live in' (Järvinen and Koskinen 2001: 10). In industrial design this is evident for example, in design theory such as sustainable and green design, wherein the design element often acts as the caring and compassionate element of a product. A rather striking recent example from Finland of the connection between design, consumption and 'care for nature' is 'Plup Water – Beverages That Clean The Baltic Sea' (http://www.plup.com), a Finnish bottled water aimed at international markets. The bottle is designed by the renowned and internationally acclaimed Finnish designer Stefan Lindfors. The strikingly designed bottle is produced in 'recyclable plastic' and 'the equivalent of 10 Euro Cents [is donated to] pollution- or disaster-related matters on the earth'. This is a product that through its design elements makes the world a better place by using recycled plastic, donating 10 cents to charities and providing an aesthetically pleasing experience to the hip yet conscious consumer – but only for as long as the consumer does not question the necessity of the commodification of public goods, such as water. Yet the product is in many ways typical of the creative economy, wherein the primacy of increased consumption is rarely questioned.

The infusion of soft characteristics is also posited to require new ways of thinking about both the economy and its subjects. Such new thinking necessitates the formation of new creative subjects who are (increasingly) creative, innovative, and entrepreneurial – yet caring. Soft capitalism increasingly manages the production and exchange of knowledge through the production of academic theories, texts and practices that embody them in subjects (Thrift 2005: 46). An illustration from a national level is the Finnish national design programme which started in 2002, with one of its main objectives being to improve the competitiveness of design education and research (Sorvali et al. 2006). The final report of the programme highlighted how the '[e]ducation of design should [be] more firmly connected as a part of technical and business education'. The most visible outcome of this thinking is the creation of the so-called Aalto University[3], which merges the Helsinki University of Technology, the Helsinki School of Economics and University of Art and Design Helsinki. The idea is that international competitiveness necessitates new designer subjects to be more entrepreneurial and this is apparent throughout Finnish policy documents on the development of the education system. The former dean of the University of Art and Design, who is one of the foremost proponents of the idea (even if the fiercest opposition has come from the students at the University of Art and Design), concludes that '[g]lobal competition has forced us to think increasingly critically about competitiveness and the foundations of sustainable 'wellbeing'. The future is seen as innovation driven, and specifically innovation ability and creativity are sources of competitiveness that all of the more developed countries strive for' (Sotamaa 2006a: 9) and '[e]ven arts has...to be linked in diverse ways to the innovation process' (11), the new creative subjects (students, the future creative workforce) will be exposed to 'a creative atmosphere, path breaking thinking, entrepreneurial spirit and intellectual stimuli' (13).

Together alone – creating creative subjects

> The road towards a more creative Finland goes through an attitudinal change and by changing the ways of thinking…. (Ministry of Education 2006: 4)

In creativity discourse the identity of (creative) workers is increasingly articulated through competition between intrastate regions and/or interstate regions. Such a competitive environment – born out of economic globalization – necessitates new citizen subjects who are entrepreneurial, flexible, innovative and now also creative (Thrift 2000; Bunnell and Coe 2005). The reason of new subject positions is anchored in the rhetoric on the radical alterations 'in the ways that cities and regions maintain competitive advantage' (Florida 2006b: 49). Florida (2002, 2005, 2006a) refers to the new subject position as 'the Creative Class'. Although a central tenet of the creative discourse is 'that everyone is creative', 'only some function socially…to create economic or cultural value from it' (Hartley 2005: 28). Florida (2002) divides the workforce into three main classes: the creative class, the service class and the working class. The creative class is supported by the serf-like service class, people with 'low-end, typically low wage and low-autonomy occupations', who exist 'mainly as a supporting infrastructure' (Florida 2002: 71) for the rapidly growing Creative Class and its consumer, leisure and work needs. The super-creative core includes 'scientists, engineers, university professors, poets and novelists, artists, entertainers, actors, designers and architects, as well as the thought leadership of modern society: non-fiction writers, editors, cultural figures, think-tank researchers, analysts and other opinion makers' (Florida 2002: 69). Designers and industrial designers are through such conceptualization seen as epitomizing the Creative Class. They are well-educated professionals who combine creativity and research to develop functional and aesthetically appealing design for new products and to satisfy increasing consumer needs, yet do it sustainably and by 'respecting nature'.

Industrial designers are not identified by industry, but through their economic function; this is a central assumption in the construction of the Creative Class. Industrial designers, much like the whole Creative Class, span a vast variety of industries and do not 'identify with the industry as a whole' (Hartley 2005: 28); an industrial designer working in the automotive industry does not necessarily identify with the automotive industry as a whole. According to Hartley (2005: 28), designers increasingly normalize a 'casual, part-time, freelance... "portfolio" career with several employers, which means that they see little common cause with each other'. Instead of identifying with an industry as a whole, the assumption is that individuals come to see themselves through a creative ethos, wherein coherence is built on 'similar tastes, desires and preferences' manifested both in work and leisure (Florida 2002: 68).

Florida (2002) refers to class not through its Marxian conceptualization, as in this 'little analytical utility remains' (68). According to Florida, in Marxian terms class is an inscribed and outdated 'basic structure of capitalists who own and control the means of production, and workers under their employ... the Creative Class do not own and control any significant property in their physical sense. Their property – which stems from their creative capacity – is an intangible because it is in their head' (Florida 2002). The implication is that workers are now in control of the means of their sustenance. This conceptualization owes much to what Bell (1973: 374) refers to as the 'Knowledge Class': 'First, the major class of the new society is primarily a professional class, based on knowledge rather than property.' Bell's conceptualization begs the question of what analytical deficits shackle knowledge as opposed to creativity. Both creativity and knowledge discourse evolves around commodification; both are portrayed as properties that can be managed and exchanged. According to Lyotard (1984: 4), '[k]nowledge is and will be produced in order to be sold, it is and will be consumed in order to be valorized in a new production: in both cases, the goal is exchange.' Creativity cannot be traded to the extent of knowledge can, as the essence of creativity is portrayed as if it cannot be codified, and thus not easily sold and bought on the market. It is inherently entwined with labour, and labour is not, as Karl Polanyi (1944) observed, produced for sale like other commodities, but rather emanates from society, that is through education and training. According to Florida (2002), the creative economy is characterized by the commodity that is inside the heads of the workers, and maintains that '...the ultimate "control" issue is not who owns the patents that may result [from creativity]' (37). However, even at the most immaterial end of the labour spectrum, intellectual property regimes govern the commodification of creativity (Thompson 2005). But traditional Marxian conceptualizations of class and labour power cannot be easily dismissed. As Thrift (2005: 47) notes regarding the subjecthoods in the new economy, the discourse reproduces a conception of the person involving both the super-exploitation of managers – who are now expected to commit their whole being to the firm, and the workers – who are expected to commit their embodied knowledge to the organization's epistemological resources without hesitation. The difference to 1930s 'organization man' is that this is achieved without bureaucratic control or hierarchical structures, instead it is achieved by government through freedom (Dean 1999; Rose 1996, 1999), wherein the subject governs him/herself by aspiring to norms (creative, caring, entrepreneurial individualists) that are both socially worthy and personally desirable (Rose 1999: 76).

Basing the subject position on creativity seemingly enables the bridging of what Bell (1976) saw as two conflicting elements, social structure, ruled by economic principles and defined by efficiency and rationality, and culture, the 'prodigal, promiscuous, dominated by an antirational, anti-intellectual temper in which the self is the measure of the aesthetic worth of experience' (Bell 1976: 37). The portrayal of these conflicting elements connects to the rebranding of cultural industries to creative industries. The shift that has occurred in terminology from 'cultural industries' to 'creative industries' marks several shifts of meaning as far as discourses of creativity are concerned. Central to this is the wrenching away of the meaning of creativity from the artistic to the public sphere. The adoption of the term 'creative industries' enables an understanding that goes 'beyond the idea that creativity is the exclusive domain of artists' (Landry 2000: xv). Such an understanding facilitates the removal of 'artificial boundaries' between culture and creative, and arts and industry. Again, 'design' is central to the discourse as '...design is one of the most dynamic and diverse of the creative industries...design is one of the key-instances of creativity...' (Cunningham 2005: 290). Industrial design, with its strong commercial and technological aspects, can still utilize the language of aesthetics and arts without appearing subversive or threatening when applied in a business setting. Competitiveness is normalized in the industrial design sector, unlike in many other cultural industries, such as performing arts or crafts. In the Finnish case, in the creation of the Aalto University a vocal critique was raised by the student union of the University of Art and Design on how, in policy documents, no other activity but industrial design at the University is highlighted or mentioned when explaining the necessity to merge the schools (TokyoWiki 2007). The critique also highlights that 'art' can be appropriated to strengthen the necessity of a merger, and at the same time art and its critique are completely marginalized as they do not fit the language of competitiveness (Hirvonen 2007).

Creativity is diffused into the spheres of technology and innovation, and this enables an understanding of creativity in relation to large parts of the ICT (Information and Communication Technologies)-sector, not least in terms of software and content-production. The relation between creativity and the non-economic enables many other tenets of life outside the economic to be now understood in relation to competitiveness – from art and education to the wellbeing of everyone. What it risks doing is creating an exclusive focus on sectors that are seen to specifically combine technology and culture, while pushing other sectors of the cultural to the margins of 'uncompetitiveness' (see Figure 3.1). Unlike art, industrial design is not pure 'art', but it requires aesthetic elements and artistic creativity (Holm 2006).

The shift from culture to creativity has also been a gateway to circumvent a number of issues connected to culture and cultural industries, most notably perhaps the position of fine arts and the critique of Theodor Adorno and Max Horkheimer of the commodification, industrialization and marketization of culture. Cultural industries are still widely understood through their relation to the Frankfurt school and the critique that cultural industries are a governing mechanism that controls the population serving market interest, and that the needs served by the cultural industries are false needs (Adorno 2001). Even when stripped of its Marxist past and even through 'morally neutral[,] use of the term "cultural industries" has proven limiting in the policy context' as it fails 'to *combine* art and culture, culture and creativity [emphasis of the original text]' (Hartley 2005: 14). Creativity will help reveal 'the realities of contemporary commercial democracies'

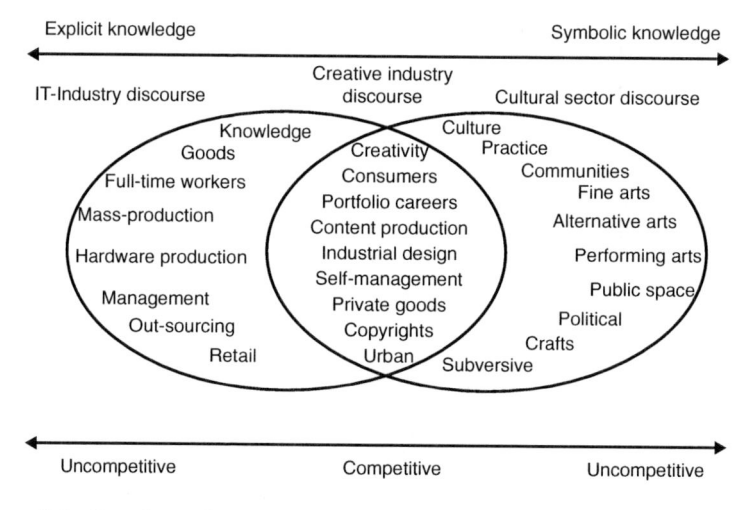

Figure 3.1 Creative industries: Empowering the centre

(Hartley 2005: 8), presumably leaving in its shadows the opposing and subversive elements of the arts domain. A recent report initiated by the European Commission ('The Economy of Culture in Europe' 2006) suggests that the creative and cultural sector should be distinguished in terms of the latter's 'cultural outputs having no secondary "utilitarian" function' ('The Economy of Culture in Europe' 2006: 45), the outputs being 'exclusively "cultural"' (53), whereas the creative sector 'gathers the remaining industries and activities that use culture as an added-value for the production of non-cultural products' (53). In the discourse it appears that designers act as a link between the cultural and commercial, as '... [d]esigners know how to turn cultural understandings into products' (Järvinen and Koskinen 2000: 15).

Designing 'Creative Finland'

From a creative economy perspective Finland has had two economic sectors that rise above others; from a historical perspective, 'cultural industry' has played an important part in establishing the Finnish knowledge economy on the international map (Stenros 2004), but during the last 20 years the ICT sector has surpassed and left the cultural industries in its wake as the spearhead of Finnish competition (Castells and Himanen 2002). Through both sectors have successfully internationalized they have had a central role in the establishment of a Finnish postwar identity, establishing a legacy and myth of 'Creative Finland'. Giving exact statistical figures on the size of either the creative economy or the industrial design sector in Finland is not easy. As Alanen (2004) notes the size of the Finnish creative industry[4] varies according to which definition is used, but estimations stand at around 35 per cent of employment (Alanen 2004), 6 per cent of the firm population and approximately 3 per cent of GDP (Ministry of Education 2008). The exact size of the Finnish design sector is also unclear, due partly to how design processes span industry borders but also due to the lack of well-defined and regularly collected statistics (Salimäki et al. 2004). Following Finnish statistical classifications, the design sector's impact on the Finnish economy is relatively small: in terms of number of firms in 2003 there were 2300 design companies, of which 1416 were architectural firms and 846 firms in industrial design (Salimäki et al. 2004). The size of the design sector grew in terms of GDP only from 0.2 in 1995 to 0.4 in 2002 (Lindström et al. 2006: 15). However, in competitiveness indices that measure the design economy, Finland usually does very well as these are not bounded by industry classifications but instead use measures such as company spending on research and development and brand value (e.g. New Zealand Institute of Economic Research (NZIER) 2003; Sorvali et al. 2006; World Economic Forum 2005).

But the design sector has always had an important role in the imagining of international competitiveness and in the creation of a national identity (Stenros 2004). In the 1950s the Finnish identity was formed by the success of individual Finnish artists and designers that ventured into the international market. Names such as Alvar Aalto (architecture, furniture and glass), Tapio Wirkkala (designer), Timo Sarpaneva (designer) and Maija Isola (printed textile design) embodied the spread of 'Finnish Design'. Finnish design was constructed internationally through the commercial success of such individuals. For ICT, the success of Nokia would have a similar effect in terms of the way Finland and Finnish culture are perceived abroad and how the Finns see themselves. Nokia in a way epitomizes the creative economy, as it is located in a small economy, commercially and internationally successful and competitive and fuses high-technology and cultural understanding into a 'design house for mobile communication' (Kotro and Panzar 2002: 36). In the 1990s, after years of decline, the notion of 'Finnish design' once again became the metaphorical heart of Finnish competitiveness.

The notion of Finnish design – what it constitutes, and what its geographical implications are – is not easily deciphered. Creativity might be less codifiable, and thus more place-based than knowledge, but it is not a national trait when analyzed through the lens of international competitiveness. The case of linking Nokia to 'Finnish design' is interesting from a geographical perspective. Nokia is selling a technological product, the mobile phone, for a global market. It acknowledges design as an important strategic 'asset' (Järvinen and Koskinen 2001: 77). The success of Nokia is often referred to in Finland as linked to the notion of (modern) Finnish design, for example, in comparisons of different countries' competitiveness in the design field, Finland often comes out on top not least due to the strength of Nokia's design brand (e.g. NZIER 2003). Research has also found that Nokia has influenced the development of Finnish design (Valtonen 2007). The linking of Nokia to Finnish design is well illustrated in the report on global competitiveness of national design sectors by the New Zealand Institute of Economic Research (NZIER 2003: 9):

> Finland's design culture has always been strong. In the 1950s and 1960s, Finnish designers and architects such as Alvar Aalto were the embodiment of 'hip modernity'. However, it has only been in the presence of a suite of national design policies that the Finnish brand

has become truly global…. The single biggest success story from Finland's design program is Nokia.

But Nokia also raises questions of if – when understood in relation to international competitiveness – the notion of 'Finnish design' does not lose its meaning, and if referring to Nokia as 'Finnish design' does not overemphasize the importance of 'global competitiveness' for the rest of the design sector, especially when referring to global competitiveness rankings to support policy decisions. Drawing strong connections between Nokia and Finnish design is contestable in many ways. Today Nokia has its main design studios in London (UK), which hosts 40 per cent of Nokia's in-house design force, Espoo (Finland), Beijing (China), and Calabassas, CA (USA), with newly opened satellite studios in Bangalore (India) and Rio (Brazil) (Nokia 2008; Exon 2007). In 1990s Frank Nuovo[5] acted as Nokia's chief designer, he was an Italian-American based at Nokia's design centre in Los Angeles, California. Currently the position is held by Alistair Curtis, who holds a Master of Arts degree in Industrial Design Engineering from the Royal College of Art in London (http://www.nokia.com). Nokia's new design phone features music composed by Grammy winning Japanese composer Ryuichi Sakamoto, who is based in New York and Tokyo. In a recent interview Curtis spoke on the reasons for having Nokia's design centre in Los Angeles during the 1990s: 'It was partly due to where Frank Nuovo was based. L.A. has always had a strong pool of design and creativity. Most if not all auto manufacturers in the world have an L.A. design studio. The speed of product development was slower in those days. You could afford to be farther away from the machine' (Curtis cited in Business Week 2006).

Today, Nokia employs about 300 designers in its global team, working from four main studios in Finland, the UK, the US and China, as well as a number of second-tier offices; it also utilizes its large network of consultancies and subcontractors for design purposes (Järvinen and Koskinen 2001: 95). In August 2007, Nokia opened its first design studio in Bangalore, India, characterized as a 'design hot-spot' and a 'design oasis' by Alistair Curtis (Business Week 2006). When speaking about Bangalore, Curtis' language mirrors that of Florida's in terms of the 'creative class', but here the description is not that of the Western-based subject, and there is no emphasis on 'Finnish' design:

> A local creative university sponsors, it's an area where we can interact with local creatives in order to sort of look at the culture of that

country. We can look at specific design issues we want to develop with them. We are a global organization and there is diversity in the marketplace. It's about being able to tap that better. What are those differences? Are the drivers the same? In many cases they are, but you may need to tweak products to make them more appealing from a Rio perspective or a Russian perspective. (Curtis cited in Business Week 2006)

Curtis also draws on the national imaginations of India as 'a dynamic and inspiring place for designers with its diverse mix of cultures, traditions and colour' (Curtis quoted in BBC 2007). These narratives both confirm and question a geographical understanding of creativity; on the one hand it shows that there might be little utility left in the notion of 'Finnish design', on the other hand it seems to conform to the understanding that creative subject positions are neither formed by nationality nor strict industrial boundaries. I thus return to the question of what actually 'holds together' the creative class.

True values of creativity

The creative class is governed and governs others through what they see as the 'true' behavioural norms and conventions of the creative economy. We can discern what behavioural norms are seen as true and valued by looking at the 'similar tastes, desires and preferences' (Florida 2002: 68) of the creative class: values that stem from the economic function of individuals. These 'mainstream' values are individuality, meritocracy, diversity and openness (Florida 2002).

Florida sees individuality as an unwillingness to 'conform to organizational or institutional directives and resist[ance to] traditional group oriented norms' (Florida 2002: 77–78). In the subjectification of the new creative worker, notions such as self-management, self-learning, and self-empowerment are central to the creative ethos. Workers prefer to 'move around' (Florida 2002: 110) and 'their only real job-security comes from their capabilities and continued productivity' (Florida 2002: 110). They accept that a significant amount of their working time is invested 'in taking care of their creativity' (Florida 2002: 114) through education, learning and picking up new skills, 'all of it unpaid' (Florida 2002: 114) during 'spare' time. Creativity is enhanced and produced for the firm but the individual worker is responsible for the production costs.

In their portfolio careers – often typical within the design sector – members of the creative class govern themselves as they 'now *expect* to manage their working lives' (Florida 2002, p. 115). 'We have come to terms with the new labor market...we simply accept [lay-offs] and go about our busy lives. We acknowledge that there is no corporation or large institution that will take care of us – that we are truly on our own' (Florida 2002: 115). Coming to terms with the new labour market ties back to the issue of labour power and apolitical values: '...the ultimate "control" issue is not who owns the patents that may result, nor is it whether the creative worker or employer holds the balance of power in labour market negotiations. While those battles swing back and forth, the ultimate control issue – the one that we have to stay focused on, individually and collectively – is how to keep stoking and tapping into the creative furnace inside each human being [sic]' (Florida 2002: 37). In terms of 'labour market negotiations', organized labour, as with most political activity, is presented as uninteresting to the creative class. By definition they span a vast variety of industries and do not 'determine the shape or structure of the social organization of creativity' as they do not 'identify with the industry as a whole' (Hartley 2005: 28). Individuality is a prime example of 'responsibilization' (Rose 1999), wherein indirect techniques are developed for leading individuals without being responsible for them. By naturalizing the values of the creative ethos, subjects become 'responsibilized' through the naturalized understanding that social risks, such as poverty, illness, housing, exclusion, illiteracy and unemployment, are not the responsibility of state protection systems but rather the responsibility of the individual. Instead such issues are represented as matters of self-care (Rose 1999; Lemke 2001), for example:

> The Creative Class values hard work, challenge and simulation.... The companies that employ them are often under tremendous competitive pressure and cannot afford much deadwood on staff: Everyone has to contribute. The pressure is more intense than ever to hire the best people regardless of race, creed, sexual preference or other factors...meritocracy ties into...values and beliefs we'd all agree are positive – from faith that virtue will be rewarded, to valuing self-determination and mistrusting rigid caste systems. (Florida 2002: 78)

In order to create a true competitive environment, which can attract the creative class, there is a need to focus on only the most 'intelligent' and 'best' subjects irrespective of their nationality or ethnicity (Bunnell and Coe 2005). By valuing only merit, meritocracy becomes inherently intertwined, and at times indistinguishable from, the value 'of openness and diversity'. The creative class should show 'tolerance for strangers and intolerance for mediocrity' (Florida 2002: 227). As 'merit' is a central value issues such as race and sexuality becomes almost a non-issue. Florida does, however, recognize that such an emphasis on meritocracy might divert from the relationship between the cultural and educational advantages that lead to 'merits', and issues on race and sexuality, but as is typical of the creative discourse these issues are conferred to the periphery. Meritocracy is thus closely connected to diversity, which is portrayed as the most fundamental value of the creative class. Diversity is singled out as the most important value that the creative class looks for in the location of the potential employer: values of openness and tolerance should thus govern the conduct of firms and local politicians. Such values are not strange to the design profession; the International Council of Societies of Industrial Design defines the task of designers as '...supporting cultural diversity despite the globalization of the world' (http://www.icsid.org). In practice signs of openness to diversity encompass inclusion of 'people of different groups and races, different ages, different sexual orientations and alternative appearances such as significant piercings or tattoos' (Florida 2002: 226).

As has been discussed in this chapter, new subject positions (the creative worker) are engineered around a number of central values, that is individuality, meritocracy, diversity and openness. From these central values, the creative class becomes characterized by its freedom: freedom of being an individual, freedom to be the best, and freedom of mind. Members of the creative class are 'governed through [their] freedom' (Rose 1999: 62), in line with the neoliberal mode of government. Once these values are naturalized and conformed to in specific places and within firms in such locales, they will accrue competitive advantages in relation to other places.

Creating 'Creative Finland'

Place is paramount in creativity discourse. Herein, I specifically focus on performative spaces, spaces where people act and try out new subject positions (Madge and O'Connor 2005). In doing this I continue to look at the establishment of a design-driven and creative economy in Finland in more general, but also use the case of Arabianranta – the City of Art and Design – as an example of performative space.

Arabianranta is an urban district in Helsinki and a major new urban development project in the metropolitan area. It is a public-private partnership wherein the City, which owns the land, gives private actors land for development, while providing infrastructure and municipal services (City of Helsinki Urban Facts 2006). The emphasis of the project lies in residential and housing development. But the most talked about and most publicized aspect about this housing project is that as a 'subproject', the City aims to establish Arabianranta as 'the leading centre of art and design by the end of the decade' (http://www.helsinkivirtualvillage.fi) and as a 'leading hub of design in the Baltic context' (Kangasoja and Schulman 2007: 17). The public private partnership is operated through a company called Art and Design City Helsinki Ltd. (ADC); its shareholders are the City of Helsinki, the Ministry of Trade and Industry, the Foundation for the Support of the University of Art and Design, the Foundation of the Pop and Jazz Conservatory, Arcada Polytechnic[6], Iittala Group and Arabian Palvelu[7].

The Arabianranta district plays a central role in the overall innovation strategy of Helsinki. The mayor of Helsinki contextualized Arabianranta in the larger competitiveness/globalization discourse when he noted that 'the goal for the Helsinki region's innovation strategy is the success of Finland and the metropolitan area's in the global competition. Arabianranta is seen in this strategy as the significant development environment for especially the creative industries' (Pajunen on http://www.helsinkivirtualvillage.fi).

The values of the creative class (individuality, meritocracy, diversity and urbanity) are acted out through practices. This relates strongly to performativity; what people do and act out is connected to the practices that reproduce discourses while enabling and disciplining subjects and their performances through 'the saturation of performers with power, with particular subject positions' (Madge and O'Connor 2005: 85). Arabianranta can be seen to be a performative space in which 'subject positions are 'tried out' in and through which practices of everyday life and these acts, or fabrications, produce over an time identity "effect"' (Madge and O'Connor 2005: 85). The identity effect alluded to is the creative worker; creativity becomes actualized through performance and through performative spaces. As subject positions are formed, the subjects start governing themselves according to the values and norms that they see as 'true' and valuable.

In the following sections I will look at some technologies and means by which creative subject positions are formed in the context of Arabianranta. I do this by looking three dimensions of the performative space of the 'City of Art and Design'. I base these three dimensions on the so-called '3Ts of economic development': Talent, Technology and Tolerance (Florida 2002). From a governmental perspective, they are part of the creation of an environment that stimulates creativity and one that would 'attract creative people, generate innovation and stimulate economic growth' (Florida 2002: 249). I thus consider how 'Talent, Technologies and Tolerance' are utilized to form creative subject positions.

Talent: How we became talented

In the creation of an environment beneficial to the design industry and favourable for creativity, it prescribes the co-presence of the most creative subjects (Bunnell and Coe 2005). In the Arabianranta/Finnish case the creation of a creative subject has been promoted in several ways. When creating an image of Finland, the history of Finnish design and creativity is often drawn upon. Names of industrial designers from the 1950s such as Alvar Aalto, Tapio Wirkkala and Timo Sarpaneva are often drawn upon to verify creativity as a national characteristic. Other historical images used for establishing a creative history include the success of Nokia, design firms such as Artek, the Finnish furniture company founded by Aalto, and Arabia, a (designed) tableware company which employed several of Finland's foremost industrial designers in the 1950s. Arabia is especially significant in the sense that it shares history with the district. Arabianranta is Finnish for 'Arabia's shore'. It was the site for Arabia's porcelain factory throughout the twentieth century and now houses the University of Art and Design.

Arabianranta is portrayed as the birthplace of Helsinki City; it is described as 'the almost mythic site of medieval Helsinki' (Bunschoten et al. 1996). Arabianranta was also the site for the first steps towards industrialization in Finland, and fostered a number of smokestacks. History is used in juxtaposition with the high technology infrastructure of the 'new' Arabianranta: 'The spirit of the place – genus-loci – is a unique weft of the synergy born out of history, wild nature as well as art, media, business and industry.' (http://www.helsinkivirtualvillage.fi). It is notable that 'genus loci' is used similarly as 'industrial atmosphere' was in the new regionalism literature. As is typical for the new economy discourse, 'natural' places and surroundings are given specific importance. The landscapes also include more active landscapes, 'landscapes of interactive lifestyle': parks, jogging trails, skate parks and so on (Bunnell and Coe 2005: 840). These play an important governmental role, in allowing relaxation from the 'mental strains' of creative work (Bunnell and Coe 2005), but also for 'self-development'. The creative class is induced to realize themselves through the active use of these landscapes of interactive lifestyle.

The postmodern architectural vision is dominated by the image of difference and diversity where modern glass-plated office buildings exist alongside historic buildings as if to foster 'a cosy ambience' (Philo and Kearns 1993: 6). Arabianranta is surrounded by the former working class quarters with low wooden buildings from the nineteenth century. The demand to preserve old buildings on the basis of cultural rather than direct economic value helps to reconcile the contradiction between market and place. The partial preservation of old building and surroundings constitutes a market for the special characteristics of place (Zukin 1991). The site was chosen as much for its metaphor as its location. The postmodern architectural vision, as used as a means of place marketing and selling places, stands in contrast to the modernist vision that ignores history and in so doing alienates people from the surroundings and from place itself (Jacobs 1974; Philo and Kearns 1993). The spirit of creativity is inscribed into the built landscape, where visions of diversity and difference are replacing the modernist homogeneous concrete buildings (Philo and Kearns 1993). This echoes the position of Jacobs (1974), whose ideas Florida claims to reproduce, in that the preservation of old buildings is a necessity to foster social diversity. As Zukin (1991: 193) shows, gentrifiers' capacity for connecting to history gives them 'license' to reclaim places for their own use.

Scattered around Arabianranta are pieces of public art. The city has stipulated that property developers have to allocate one to two per cent of the acquisition costs for artistic projects. The 'theme' of these projects is to be based on the 'area's history and nature' (http://www.helsinkivirtualvillage.com). People are constantly exposed to the artistic element of Arabianranta. Such place-based association with the arts invests the neighbourhood and new developments with an 'aura of authentic cultural consumption' (Zukin 1991: 192), and more importantly in Arabianranta's case, production. The public art in Arabianranta is seen as 'important for forming the identity of the area' (http://www.uiah.fi/). It is equally important for forming the subject positions and identities. Of course it visualizes the idea that creative industries include not only the private but subsidized public art. People who are potential dwellers of Arabianranta are portrayed as becoming part of the creative class through location. It can also be understood as participating in modes of self-government based on lifestyle choices (Bunnell and Coe 2005). The relocation to Arabianranta is not, however, only about art or real estate; it is also about getting access to the frontier of information technology. This brings us to the second dimension of performative space, technology and virtual space.

Technology: Helsinki Virtual Village

One of the aims of the Arabianranta region is to create a resident-centred, communal urban area, which communicates in a versatile way with the help of technology. The area can act as a national test bed for experimental projects in the creative industries. (Eskelinen 2005: 6)

The physical performative space of Arabianranta is mirrored by a (cyber) space of performativity, Helsinki Virtual Village (HVV), a password-protected web forum for residents only. All the buildings and apartments in Arabianranta are equipped with state of the art fibre optic networks and wireless infrastructure. By having every building connected to the global networks and providing wireless Internet access around the neighbourhood the boundaries between work and leisure become blurred. Florida reflects on this by describing how the new suburban high-tech campuses have 'virtually everything a worker would want or need – from espresso bars and free food to on-site day care, state of the art health facilities, outdoor Frisbee fields and concierge services. The message and function are clear: "no need to go wandering off; stay right here at work"' (Florida 2002: 123).

HVV can also be seen as a performative space, as subject positions can also be tried out and through practices of everyday life. These (cyber) subjectivities may then spill over to place-based identities either by supporting, blending in or by undermining it (Madge and O'Connor 2005). According to one resident '... as soon as the discussions started on [HVV] I started to identify with the new place... I started to feel that I belong to the house, and also to the place as a whole' (Kangasoja 2007: 151). For others, the (cyber) subject position is felt as clashing and undermining their identities. The new subject position, mediated through constant exposure to technological possibilities and images of an internet savvy, creative, fast subject, was resisted by some of the residents. Some people felt inadequate and anxious as society imposed these new technological identities on to them (Kangasoja 2007).

Living Lab is another performative (cyber) space of Arabianranta. The idea is that the residents are using and testing prototypes and products in their homes as well as providing information to companies about their lifestyles over the local network: 'The Helsinki Living Lab concept aims to be as close as possible to the end user and the use of the service in everyday life and in genuine conditions of normal use. As a result we can create products and services that people truly consider necessary, useful and positive' (http://www.helsinkilivinglab.fi). Here, although the participants are voluntary 'producers', instead of consumers, the residents do not, of course, choose what they produce. They are being told what to consume and experience (Kangasoja 2007). HVV and Living Labs are not platforms for networked democracies; rather, they are technologies that strengthen subject positions, governing how inhabitants see Arabianranta, and ultimately themselves and others.

Tolerance: Not in my neighbourhood

> The places that are open and tolerant – the places where gays, bohemians and immigrants feel at home and where there is greater racial integration – tend to have a culture of tolerance and open-mindedness.... Regions and nations that have such ecosystems... gain a tremendous competitive advantage. (Florida and Tinagli 2004: 25–6)

Tolerance is posited to be perhaps the most central value of the creative class. This is something of a conundrum for the Finnish self-image. They might see themselves as talented and technologically savvy, but how about tolerant? Finland is a country with a fairly homogenous population compared to other EU countries; the proportion of immigrants is very low in Finland. At the end of 2002, the number of foreigners amounted to 1.99 per cent of the population. A recent study had Finns looking at themselves through numbers that stated that 78 per cent of the population do not approve of alcoholics as neighbours, 44 per cent do not want people with mental illnesses as neighbours, one-third do not want skate parks in their neighbourhood, and about 20 per cent do not want people who have AIDS, who are homosexual or immigrants as their neighbours (Suomen Gallup 2007). Yet, creativity indices point to the tolerance of Finns and is not creativity a sign of tolerance? Florida (2004) suggests that tolerance might be a latent value in Finns: 'to realize this latent advantage that stems for their underlying attitudes and values, these nations [such as Finland] will have to liberalize their immigration policies and become more open to talent from around the world.' Diversity and openness is often reproduced in strategy and policy documents: 'In order to attract and keep creative talents, the Helsinki Region will have to provide creative settings offering high standards in housing, work and leisure opportunities.... The entire Helsinki Metropolitan Area must invest in diversified, pluralistic and increasingly international cultural provision so that foreigners will also come to appreciate that the Region provides a satisfying living environment' (Innovation Strategy Helsinki Metropolitan Area 2005: 22). But the population of Arabianranta is selectively uniform.

The inhabitants see themselves as 'predominantly families with children, generally affluent people...homogenous rather than representing urban diversity...youthful, middle-aged, relatively highly educated middle class' (Mäenpää 2007). Only three per cent of the population are unemployed, compared to ten per cent overall for Helsinki.

Perhaps Arabianranta is most clearly defined in terms of what it is not? The Principal of the Polytechnic states, regarding the school's decision to locate in Arabianranta: 'In choosing a new campus it was important that the new place would interact with its environment. We were looking for an open and extrovert campus, not some kind of ghetto' (Wolff in Rönkä 2007: 249). With this concluding thought, there is reason further to discuss what the role of economic geography research should be: is it demoted to scientifically identifying which spaces are ghettos and which are suitable for creative gentrification... or is it something else?

Conclusion

In this chapter I have considered the subject positions that are made available and valuable to us in creativity discourse and the role of industrial design in achieving the establishment of a new creative ethos. In Finland a central tenet in creativity discourse is that creativity – and industrial design – is inherently good as it represents the subversive, competitive, public, private, elitist, communalist, consumerist and sustainable, all at the same time and apparently without contradiction. The values of the Creative Class, meritocracy, openness and individuality, are portrayed as benevolent and good. Why are these values valuable to government? Because they lead us to alienate ourselves from the need of social protection systems, accepting more flexible working conditions (for example, by glorifying the flexible 'portfolio career' of designers) with less employment security, and to take care of the development of the means of production (our creativity) ourselves and without compensation – all seemingly by our own will. The increased acceptance of the creative ethos can be seen as an example of government through freedom (Rose 1999), through people's will to enact their lives in terms of activity, enterprise, choice and creativity.

Space plays an important role in forming and stabilizing these positions. Creativity discourse enables less emphasis to be placed on national policies towards regional spaces, spatial selectivity and 'splintered' urban spaces (Bunnell and Coe 2005: 834) and 'of privatized, highly networked global hotspots'. These spaces can be seen as performative spaces, space where subject positions are acted out, where the 'right' values are constituted. Research both within economic geography and organization studies can play an important role in exploring these spaces and their role in establishing the creative economy, but research needs to move beyond glorifying certain urban creative hotspots, and window-dressing them in attractive shrouds for the policymakers to admire and reproduce. Creativity is an alluring signifier to place on the centre stage of discourses that are created by economic geographers. Together with the city-regions paradigm it becomes a forceful export into other disciplines and a persuading sell to policymakers. The urban creativity paradigm promises to deliver where the hoards of cluster, industrial district and learning region studies might be perceived to have failed. The focus on the individual expands from the purely economic towards what seems a more people-centred approach; similarly when individuality is positioned as a central value it diverts attention away from the previous emphasis on trying to find evidence of local cooperation and strong local ties. Creativity – and its core producer industrial design – is like motherhood and apple pie; it becomes almost impossible to be against and criticize. Industrial design when fuelled by creativity makes the world a 'world a better, a more beautiful, a safer, and a smoother place to live in' (Järvinen and Koskinen 2001: 10). Yet when understood increasingly and exclusively in relation to international competitiveness, it becomes both persuasive and paradoxical. We need to focus on the paradoxes, and explore the silences that creativity discourse produces.

Examples of silences that we risk reproducing include issues about contradictions between the consumerist lifestyle of the creative class and issues of ecological and social sustainability amidst claims that the creative class is truly caring – as if access to parks or a design aspect to recyclable water bottles ensured ecological stability. Other contradictions in the formation of subject positions include the increasingly precarious working conditions of the creative class combined with minimization of interest in labour relations and politics. And perhaps the biggest elusive economic issue of all, that of uneven development, remains left to its own device. These issues should not be demoted to the asterisks of research and suggestions for further research. The creative economy implies little shift if the subversive elements of creativity discourse are downplayed, and the dark side of best-in-class regions are silenced or selectively reported in the hunt for a new economic orthodoxy.

I have also discussed the notion of performative space, and how space can be used to form new subject positions. The question remains: can these imposed identities and values be resisted, and are there alternative imaginations on the horizon? I will end on an inkling of a positive note, with an example that this may be possible from the Arabianranta case. Arabianranta had been marketed as a high tech, creative and caring place and one that is child-friendly to boot. The advertisements had maps with day care centres 'in nearly every yard' (Ho 2007: 236). Yet in 2005, several years after the completion of the neighbourhood, there was still an urgent need for care centres, because of 'incorrect demographic estimates' (Ho 2007). Residents were informed that a day care centre was in the pipeline for 2008. As dissatisfaction reached its peak the residents organized a survey via HVV, and in a matter of weeks the number of children living in the area and in need of day care was calculated. This was then used in negotiations with the City and in media discussions (Kangasoja 2007). The image of the inhabitants, in this case mothers using HVV this way might not have been an illustration of the imagined industrial designer inhabiting the City of Arts and Design and drawing on its genus loci to gain competitive advantages or produce more products for us to enjoy, but it can be seen as a competing imagination of what constitutes a creative class. Given, this might be seen as yet another carefully selected, feel-good detail much like that which I have previously critiqued. But it is also an illustration of how appropriation works both ways, and how places can matter in resisting the imposed subject positions. Yet these are not the stories and the voices of resistance that we sell to policymakers. Instead, if we do not stop to contemplate the values and images of landscapes that we reproduce, we stand in danger of silencing the very voices we claim to help by reproducing the creativity discourse.

Notes

1. For a critique of the creativity discourse see for example, Thrift 2005; Peck 2005; Jeanes 2006.
2. The strategy was eventually not included in the new government's policy portfolio in 2007.
3. The name obviously draws on the images and myth around Finland's most famous international designer and architect, Alvar Aalto.
4. Includes advertising, architecture, arts, crafts, design, cinema and video production, music, other performing arts, publishing, programming and radio and TV production.
5. Prior to joining Nokia as head designer Nuovo had worked 10 years for BMW Group DesignworksUSA in Los Angeles.
6. A homeware design company that owns the Arabia brand.
7. The area's property service company.

References

Adorn, T.W. 2001. 'Culture industry reconsidered'. In Bernstein, J.M. (ed.) *The culture industry – Selected essays on mass culture*. 98–107. Routledge, London.

Alanen, A. 2004. 'Luovuuden talous haussa (Looking for the creative economy)'. *Tietoaika* 1.

Bell, D. 1976. *The cultural contradictions of capitalism*. Basic Books, New York, NY.

Bell, D. 1973. *The coming of post-industrial society: A venture in social forecasting*. Basic Books, New York, NY.

Bilton, C. 2007. *Management and creativity: From creative industries to creative management*. Blackwell Publishing, Oxford.

Brenner, N. 1999. 'Globalisation as reterritorialisation: The re-scaling of urban governance in the European Union', *Urban Studies* 36:3, 431–51.

Brenner, N. 2004. *New state spaces: Urban governance and the rescaling of statehood*. Oxford University Press, New York, NY.

Bristow, G. 2005. 'Everyone's a "winner": problematising the discourse of regional competitiveness', *Journal of Economic Geography* 5:3, 285–304.

Bunnell, T. and Coe, N. 2005. 'Re-fragmenting the "political": Globalization, governmentality and Malaysia's multimedia super corridor', *Political Geography* 24:7, 831–49.

Bunschoten, R., Hasdell, P. and Hoshino, T. 1996. 'Arabianranta, Helsinki'. In Verwijnen, J. and Lehtivuori, P. (eds) *Managing urban change*. 153–61. University of Art and Design Helsinki UIAH, Helsinki.

Business Week 2006. 'Online extra: A chat with Nokia's Alastair Curtis'. 17 July. (http://www.businessweek.com/magazine/content/06_29/b3993071.htm).

Cairncross, F. 1997. *The death of distance: How the communications revolution is changing our lives*. Orion, London.

Camagni, R. 1991. 'Introduction: from the local "milieu" to innovation through cooperation networks'. In Camagni, R. (ed.) *Innovation networks: Spatial perspectives*. 1–9. Belhaven Press, London.

Castells, M. 1996. *The rise of the network society*. Blackwell, Cambridge, MA.

Castells, M. and Himanen, P. 2002. *The information society and the welfare state: the Finnish model*. Oxford University Press, Oxford.

City of Helsinki Urban Facts 2006. Helsingin kaupungin talous- ja suunnittelukeskus, Helsingin kaupungin tietokeskus, City of Helsinki: Helsinki

Cunningham S. 2005. 'Creative enterprises'. In Hartley, J. (ed.) *2004 Creative industries*. 282–98. Blackwell Publishing, Oxford.

Dean, M. 1999. *Governmentality: Power and rule in modern society*. Sage, London.

Desrochers, P. 2002. 'Local diversity, human creativity, and technological innovation', *Growth and Change* 32:3, 369–94.

Dicken, P. 2007. *Global shift: Mapping the changing contoura of the world economy*, 5th edition. Sage, London.

Drake, G. 2003. '"This place gives me space": place and creativity in the creative industries', *Geoforum* 34:4, 511–24.

Drucker, P. 1993. *Post-capitalist society*. Butterworth Heinemann, Oxford.

Eskelinen, J.-E. 2005. *Media Centre LUME 5 years*, Media Centre Lume, University of Art and Design Helsinki.

Exon, M. 2007. 'Delaney quits plan for Nokia', *DesignWeek*, 16 May. (http://www.designweek.co.uk/Articles/134811/Delaney+quits+Plan+for+Nokia.html).

Fleming, P. and Spicer, A. (2008). 'Beyond power and resistance: New approaches to organizational politics', *Management Communication Quarterly* 21:3, 301–09.
Florida, R. 2002. *The rise of the Creative Class: And how it's transforming work, leisure, community and everyday life.* Basic Books, New York, NY.
Florida, R. 2005. *Cities and the Creative Class.* Routledge, New York, NY.
Florida, R. 2006a. 'Lecture in Savannah', Speech 83 min., 47 sec., 14 December, audio recording available at http://www.evoca.com/user_profile.jsp?uid=11459.
Florida, R. 2006b. 'The future of the American workforce in the Global Creative Economy', *Cato Unbound*, 4 June. (http://www.cato-unbound.org/2006/06/04/richard-florida/the-future-of-the-american-workforce-in-the-global-creative-economy/.
Florida, R. and Tinagli, I. 2004. *Europe in the Creative Age.* Demos, London.
Friedmann, J. 1986. 'The World City Hypothesis', *Development and Change* 1, 69–84.
Gibb, S. and Waight, C.-L. 2005. 'Connecting HRD and creativity: From fragmentary insights to strategic significance', *Advances in Developing Human Resources* 7:2, 271–86.
Gibson, C. and Klocker, N. 2004. 'Academic publishing as "Creative" Industry, and recent discourses of "Creative" Economies: Some critical reflexions', *Area* 35:4, 423–34.
Hartley, J. 2005. 'Creative identities'. In Hartley, J. (ed.) *2004 Creative industries.* 1–40. Blackwell Publishing, Oxford.
Heelas, P. 2002. 'Work ethics, soft capitalism and the "turn to life"'. In Du Gay, P. and Pryke, M. (eds) *Cultural economy.* 78–96. Sage, London.
Hirvonen, E. 2007. 'Rehtori Sotamaalla on unelma (Rector Sotamaa has a dream)', *Vihreä Lanka*, 30 January. http://www.vihrealanka.fi/node/896 (accessed March 2008).
Ho, Y. 2007. 'Arabianranta in Helsingin Sanomat'. In Kangasoja, J. (ed.) *Arabianranta – Rethinking urban living.* 230–38. City of Helsinki Urban Facts, Helsinki.
Holm, I. 2006. *Ideas and beliefs in architecture and industrial design*, Doctoral thesis, Oslo School of Architecture and Design, Oslo.
ICSID (http://www.icsid.org/about/about/articles31.htm) (accessed October 2008).
Imagine a Toronto... Strategies for a Creative City, Strategies for Creative Cities Project Team, 2006-07-24. (http://www.web.net/~imagineatoronto/fullReport.pdf) (accessed May 2007).
Innovation Strategy – Helsinki Metropolitan Area, Culminatum, 2005. (http://www.helsinkiregion.com/) (accessed June 2007).
Jacobs, J. 1974. *The death and life of great American cities: The failure of town planning* (reprint, first published in 1961 ed.). Pelican Books, Middlesex.
Järvinen, J. and Koski, E. 2006. *Nordic Baltic Innovation Platform for Creative Industries*, Nordic Innovation Centre, Oslo.
Järvinen, J. and Koskinen, I. 2001. *Industrial design as a culturally reflexive activity in manufacturing.* Publication series of the University of Art and Design Helsinki A 33, Helsinki.
Jeanes, E.-L. 2006. '"Resisting Creativity, Creating the New". A Deleuzian Perspective on Creativity', *Creativity and Innovation Management* 15:2, 127–34.

Jones, D. and Boon, B. 2006. 'The voice of the Creative Economy'. *Critical Management Studies Conference 2007 (Manchester, 1113 July 2007)*, Electronic Journal of Radical Organization Theory.

Kangasoja, J. 2007. 'From virtual visions to everyday services'. In Kangasoja, J. (ed.) *Arabianranta – Rethinking urban living*. 142–58. City of Helsinki Urban Facts, Helsinki.

Kangasoja, J. and Schulman, H. 2007. 'Introduction'. In Kangasoja, J. (ed.) *Arabianranta – Rethinking urban living*. 16–20. City of Helsinki Urban Facts, Helsinki.

KEA European Affairs 2006. *Economy of Culture in Europe*, report commissioned by the European Commission, DG EAC 03/05, available at: http://ec.europa.eu/culture/key-documents/doc873_en.htm, KEA European Affairs, Brussels.

Koivunen, H. 2004. *Onko kulttuurilla vientiä? (Are there exports for culture?)*, Publications of the Ministry of Education, 22.

Kotro, T. and Pantzar, M. 2002. 'Product development and changing cultural landscapes – Is our future in "snowboarding"?', *Design Issues* 18:2, 30–45.

Kulkki, S. 2004. 'Towards an ideapolis: The creative Helsinki Region', *City of Helsinki Urban Facts Quarterly Publication* 3, 18–21.

Landry, C. 2000. *The Creative City: A toolkit for urban innovators*. Earthscan Publications, London.

Lemke, T. 2001. ' "The birth of bio-politics": Michel Foucault's lecture at the Collège de France on neo-liberal governmentality', *Economy and Society* 30:2, 190–207.

Lindström, M., Nyberg, M. and Ylä-Anttila, P. 2006. *Ei vain muodon vuksi-Muotoilu on kilpailuetu (Not only because of form-design as a competitive advantage)*, the Research Institute of the Finnish Economy – ETLA (B220-series), Helsinki.

Lyotard, J.F. 1984. *Post-modern condition: A report on knowledge*. University of Minnesota Press, Minneapolis, MN.

Madge, C. and O'Connor, H. 2005. 'Mothers in the making? Exploring liminality in cyber/space', *Transactions of the Institute of British Geographers* 1, 83–97.

Mäenpää, P. 2007. ' "Residents" experience of Arabianranta'. In Kangasoja, J. (ed.) *Arabianranta – Rethinking urban living*. 170–207. City of Helsinki Urban Facts, Helsinki.

Ministry of Education 2004. *Luovuuskertomus. Ehdotus hallitusohjelmassa tarkoitetun luovuusstrategian tekemisen luonteesta, lähtökohdista ja toteuttamisen tavoista (The creativity report – Proposition for a creative strategy, its nature, base, and implentation)*, Opetusministeriön julkaisuja/Publications of the Ministry of Education 4.

Ministry of Education 2006. *Yksitoista askelta luovaan Suomeen; Luovuusstrategian loppuraportti (Eleven steps to a creative Finland: Final report on the creativity strategy)*, Opetusministeriön julkaisuja/Publications of the Ministry of Education 43.

Ministry of Education 2008. *Näin suomalaista kulttuuria viedään: Kulttuurivientiraportti 2007 ja esitykset kehittämistoimenpiteiksi (The export of Finnish culture: A report on cultural exports and suggestions for development)*, Opetusministeriön julkaisuja/Publications of the Ministry of Education 15.

Nokia. 2008. *Nokia opens satellite design studio in Rio de Janeiro*, Press release, 27 February.

NZIER 2003. *Building a case for added value through design*, February 2003, NZ Institute of Economic Research, Wellington (NZ).

O'Brien, R. 1992. *Global financial integration: The end of geography*. Pinter, London.
Oinas, P. 2005. 'Finland: A success story?', *European Planning Studies* 13:8, 1227–44.
Pajunen, J. (sine anno). 'Arabianrantaan kohoaa ainutlaatuinen osaamis – ja elämiskeskus (In Arabianranta a unique knowledge and experience center emerges)', available at http://www.helsinkivirtualvillage.fi (accessed October 2006).
Peck, J. 2005. 'Struggling with the Creative Class', *International Journal of Urban and Regional Research* 4, 740–70.
Philo, C. and Kearns, G. 1993. 'Culture, history, capital: A critical introduction to the selling of Places'. In Kearns, G. and Philo, C. (eds) *Selling places – The city as cultural capital, past and present*. 1–33. Pergamon Press, Oxford.
Piore, M.J. and Sabel, C.F. 1984. *The second industrial divide: possibilities for prosperity*. Basic Books, New York, NY.
Polanyi, K. 1944/2001. *The great transformation: the political and economic origins of our time*. 2nd edition. Beacon Press, Boston, MA.
Porter, M. 1990. *The competitive advantage of nations*. MacMillan, London.
Porter, M. 2000. 'Location, competition, and economic development: Local clusters in a global economy', *Economic Development Quarterly* 14:1, 15–34.
Power, D. and Hallencreutz, D. 2004. 'Profiting from creativity? The music industry in Stockholm, Sweden and Kingston, Jamaica'. In Power, D. and Scott, A.J. (eds) *Cultural industries and the production of culture*. 224–42. Routledge, London.
Ray, L. and Sayer, A. 1999. 'Introduction'. In Ray, L. and Sayer, A. (eds) *Culture and economy after the cultural turn*. 3–24. Sage, London.
Rönkä, K. 2007. 'The Kumpula-Arabianranta city campus'. In Kangasoja, J. (ed.) *Arabianranta – Rethinking urban living*. 246–61. City of Helsinki Urban Facts, Helsinki.
Rose, N. 1990. *Governing the soul: The shaping of the private self*. Routledge, London.
Rose, N. 1996. *Inventing our selves*. Cambridge University Press, Cambridge.
Rose, N. 1999. *Powers of freedom: Reframing political thought*. Cambridge University Press, Cambridge.
Salimäki, M., Ainamo, A. and Salmenhaara, K. 2004. *Country report: Finnish design industry*, Report prepared for the research project: The future in design: the competitiveness and industrial dynamics of the Nordic design industry, Nordic Innovation Centre, Oslo.
Sassen, S. 1991. *The global city: New York, London, Tokyo*. Princeton University Press, Princeton, NJ.
Schoenberger, E. 1998. 'Discourse and practice in human geography', *Progress in Human Geography* 22:1, 1–14.
Scott, A.J. 2000. *The cultural economy in cities*. Sage Publications, London.
Scott, A.J. 2004. 'Cultural-products industries and urban economic development: Prospects for growth and market contestation in global context', *Urban Affairs Review* 39, 461.
Sorvali, K., Hytönen J. and Nieminen, E. 2006. *Global design watch*, April, Designium, Helsinki.
Sotamaa, Y. 2006a. 'Design, innovaatioyliopisto ja luova yhteiskunta (Design, the Innovation University and the creative society)', *Chydenius-OP Ryhmän Talousjulkaisu* 3, 9–14.

Sotamaa, Y. 2006b. 'Ethics and the global responsibility'. In Salmi, E. and Anusionwu, L. (eds.) *Cumulus Working Papers: Nantes*, 16/06, 5–7. Publication Series G, University of Art and Design Helsinki.

Stenros, A. 2004. 'The story of Finnish Design', *Virtual Finland*, March. (http://virtual.finland.fi/), (accessed July 2007).

ten Bos, R. 2006. 'The ethics of business communities'. In Clegg, S.R. and Rhodes, C. (eds) *Management ethics: Contemporary contexts*. 13–31. Routledge, London.

Thompson, P. 2005. 'Foundation and empire: A critique of Hardt and Negri', *Capital & Class* 86 (Summer 2005), 73–98.

Thrift, N. 2000. 'Performing cultures in the new economy', *Annals of the Association of American Geographers* 4, 674–92.

Thrift, N. 2005. *Knowing capitalism, (Theory, culture and society)*. Sage, London.

TokyoWiki 2007. 'Homepages of TOKYO the student union of the University of Arts and Design, "Muutosehdotukset lausontoon huippuyliopistosta" (Suggestions of change to the report on the Innovation University)'. (http://wiki.uiah.fi/Tokyo/index.php/Muutosehdotukset_lausuntoon_huippuyliopstosta) (accessed September 2008).

Törnqvist, G. 2004. *Kreativitetens Geografi (The geography of creativity)*. SNS Förlag, Stockholm.

UNCTAD 2008. *The creative economy – The challenge of assessing the creative economy towards informed policy-making*, UNCTAD/DITC/2008/2. (http://www.unctad.org/creative-economy).

Valtonen, A. 2007. *Redefining industrial design – Changes in the design practice in Finland*, Ph.D. thesis, Publication series of University of Art and Design Helsinki A 74, Helsinki.

Wilson, N.C. and Stokes, D. 2005. 'Managing creativity and innovation: The challenge for cultural entrepreneurs', *Journal of Small Business and Enterprise Development* 3, 366–78.

World Economic Forum 2005. *The Global Competitiveness Report 2005/2006*. Palgrave MacMillan, Basingstoke.

Zukin, S. 1991. *Landscapes of power – From Detroit to Disney World*. University of California Press Ltd., London, England.

4
Locational Patterns and Competitive Characteristics of Industrial Design Firms in the United States

Alan D. MacPherson and Vida Vanchan

Introduction

The U.S. industrial design sector accounts for only a tiny fragment of domestic producer service employment, yet its importance to the U.S. manufacturing base is arguably disproportionate to its size. With only around 12,000 workers spread across 1800+ establishments, most firms in this sector are micro-businesses that employ fewer than ten people (U.S. Census Bureau 2007). Few consultancies have more than one business location, and most are single-person units. According to the latest data from the U.S. Census Bureau, around 73 per cent of U.S. industrial design companies in 2005 had between one and four employees.

Industrial design firms help improve their clients' revenues, product quality, styles and aesthetics, technological performance, and ergonomics (Vanchan 2007). In addition, such firms help reduce their clients' product defect rates and ease their manufacturing performance (Vanchan 2007). As a result, they offer innovative services that are inarguably beneficial to their clients as well as to the U.S. economy. The aim of this chapter is to sketch the geography of design specialization across the U.S., paying particular attention to the competitive characteristics of firms in this industry. A notable feature of the U.S. industrial design sector is that most firms are less than 20 years old. Firms in this industry tend to serve clients in a single sector, and few venture beyond their core market segments. Although the design sector has not been growing in terms of employment, new firms continue to enter this market on a steady basis.

Our chapter addresses a number of themes that have recently emerged in the literature on advanced producer services. First, specialist firms tend to compete on the basis of reputation, service quality, referrals, and trust-intensive relationships with longstanding clients (Beyers 2005). The available evidence suggests that cost-driven externalization has never been a significant factor in the use of external producer services by client firms. Instead, clients typically seek new or innovative services that cannot easily be replicated via in-house effort (Wood 2005). A second theme is that design contracts have become increasingly performance-linked (i.e. risk-sharing agreements are common). Third, the industry is gradually moving towards research-based activities such as technological forecasting and design tradeoff analysis. Design Tradeoff Analysis (DTA) allows clients to statistically evaluate potentially competing design objectives such as product performance versus ease of manufacture. Fourth, some of the more successful firms in this industry belong to elaborate innovation networks that connect different types of specialists. The presence of these networks explains why so many micro-businesses in this sector can offer a full range of design services. Finally, the trend toward design outsourcing by U.S. manufacturing companies suggests that design consultancies face bright prospects in terms of future workloads.

Like other segments of the advanced producer services, the design industry tends to agglomerate in major metropolitan areas – giving rise to high levels of spatial concentration that could intensify existing regional economic inequalities (Daniels and Bryson 2005). The design industry has also become increasingly export-oriented (Vanchan and MacPherson 2008), despite the contact-intensive nature of the service delivery process (discussed later). Prior to a discussion of these themes, however, it is pertinent to establish a geographical context for the analysis. Where are these design specialists located? And what do they do?

Geographical context

The U.S. industrial design sector is spatially concentrated in a small number of states, with the top ten states accounting for over 70 per cent of total employment. California leads the nation (20.8 per cent of total employment), followed by Michigan (10.1 per cent), New York (9.6 per cent), Pennsylvania (6.2 per cent), and Oregon (6.1 per cent). The sector as a whole consists of 1,862 firms, with total employment standing at 11,968 in 2005 (see Table 4.1). On average, U.S. design consultancies employ 6.4 workers – an average that varies little by state. An

Table 4.1 Top ten states in terms of employment counts (2005)

Rank	State	Employment	% of total	Average firm size
1	California	2494	20.8	7.1
2	Michigan	1210	10.1	13.6
3	New York	1155	9.6	6.5
4	Pennsylvania	750	6.2	6.7
5	Oregon	732	6.1	7.2
6	Texas	542	4.5	6.9
7	Ohio	533	4.4	8.1
8	New Jersey	515	4.3	6.3
9	Florida	427	3.6	6.4
10	Massachusetts	422	3.5	6.5
Sub-total		8780	73.2	Mean = 7.3
Total		11986	100.0	Mean = 6.8

Data source: American FactFinder, County Business Patterns, and Economic Census, U.S. Census Bureau (2007)

exception is Michigan, where the mean employment size (13.6 workers) is more than double the national average. This largely reflects the fact that Michigan is the nation's core region in terms of automotive and machine tool production. Design consultancies located in the Detroit metropolitan area are typically much older than their counterparts from other U.S. regions.

Table 4.2 lists the top ten metropolitan areas in terms of their percentage share of design service establishments. These rankings are virtually identical when employment ranges are used. The rankings are compared with population shares for each metropolitan area. In almost every case, the top ten metropolitan areas have a share of design service activity that significantly exceeds their population share. In short, the metropolitan distribution of design service activity is not simply a function of city size. Overall, the top ten cities enjoy a 42.8 per cent share of national design service activity, compared with only a 23.4 per cent share of the nation's total population count. For all but one of the metropolitan areas listed, establishment-based location quotients for design services exceed 1.5 (Philadelphia is the exception, with a location quotient of 1.1). There is a temporal consistency to these patterns, in that the 1997 rankings are identical.

Keeping these points in mind, there is a noticeable geography of design service specialization across the nation. Specifically, the types of services offered in any given metropolitan centre tend to mirror the structure of local production (i.e. prominent or dominant sectors).

84 *Alan D. MacPherson and Vida Vanchan*

Table 4.2 Top ten metropolitan areas (establishment counts) and metropolitan population sizes (2005)

Rank	Metropolitan area	% of design establishments	% of population
1	New York City	10.69	6.35
2	Los Angeles	9.13	4.36
3	Miami	4.99	1.83
4	Chicago	4.40	3.19
5	San Francisco	3.38	1.40
6	Boston	2.42	1.50
7	Detroit	2.42	1.51
8	Philadelphia	1.83	1.96
9	San Jose	1.83	0.59
10	Portland (OR)	1.77	0.71
Total		42.86	23.40

Data Source: Economic Census and American FactFinder, U.S. Bureau of Census (2007)

Bryson and Rusten (2005) found a similar trend on the regional geography of Norwegian knowledge intensive firms, where there is a strong concentration of these firms in the major cities. Table 4.3 lists the top ten metropolitan areas in terms of establishment counts and market specializations (note that only figures on employment ranges were given at the metropolitan scale due to disclosure restrictions). The lack of hard employment data is not a major problem, if only because similar rankings emerge when establishment counts are compared with employment ranges. New York City leads the nation as a whole, followed by Los Angeles, Miami, Chicago, and San Francisco.

For each metropolitan area listed in Table 4.3, we conducted internet-based searches to inspect the websites of resident design consultancies. We then allocated individual firms to their core market segments on the basis of self-reported specializations. In the case of New York City, for example, we are able to locate 104 of the 199 listed establishments via internet searching. Sixty-three of these 104 companies were clearly focused on apparel or textile design, with 35 having business locations within a one-mile radius of the city's garment district. Repeating these procedures for the remaining metropolitan areas yielded a regional pattern of design specialization that closely matches the nature of local industrial activity. While there is no doubt that industrial designers are a subset of Florida's (2002) creative class, their spatial distribution does not correlate strongly with any kind of 'coolness index'. As a result, design-related activity tends to mirror the geography of

Table 4.3 Top ten metropolitan areas: Establishments and market specializations (2005)

Rank	Metropolitan area	Establishment count	Market specialization	%
1	New York City	199	Apparel/textiles	62
2	Los Angeles	170	Electronics	58
3	Miami	93	Household goods	52
4	Chicago	82	Machinery	54
5	San Francisco	63	Aerospace	50
6	Boston	45	Electronics	66
7	Detroit	45	Auto-parts/machinery	76
8	Philadelphia	34	Household goods	50
9	San Jose	34	Electronics	65
10	Portland (OR)	33	Electronics/aerospace	60

Data Source: Market specialization figures were compiled by the authors and the establishment counts are from Economic Census and American FactFinder, U.S. Bureau of Census (2007)

production (which includes both high- and low-technology industries). For example, over 60 per cent of the design consultancies in the Los Angeles area are focused on electronics or aerospace markets, 76 per cent of Detroit's design companies serve the automotive or machinery sectors, and half of San Francisco's design companies cater to clients in the aerospace industry. In addition, 'cool' cities such as Seattle and Los Angeles certainly have an above average share of design companies that serve clients in the aerospace sector. On the other hand, somewhat 'less cool' cities such as Detroit and Buffalo have an above average share of design companies that serve clients in the automotive and machinery sectors. Although each of the top ten metropolitan areas contains design companies that cater to other sectors as well (notably household products), there would appear to be a high level of specialization by city. This is further evidenced in the case of specific cities outside the top ten. A prominent example is Seattle, where over 70 per cent of resident design consultancies serve the commercial aerospace market (Boeing is the ultimate consumer in this example).

Although total employment in the design sector has remained relatively stable over the last ten years (averaging 11,000 workers), there has been a trend toward the emergence of micro-businesses with fewer than five employees. In California, for instance, 116 new companies entered the design market between 1997 and 2005, cutting the average firm size from 8.9 to 7.1 workers. While this is not a dramatic drop, the same general trend has been taking place across the nation as a whole. Specifically, the population is shifting toward a youthful age structure. By now, in fact, over 70 per cent of the design consultancy population consists of micro-businesses that are less than 20 years old.

On this note, the remainder of our chapter summarizes the results of two recent firm-level surveys that were conducted to shed light on the organizational and competitive characteristics of design service vendors and their clients. The first survey was conducted in 2005, and the results are reported in Vanchan and MacPherson (2008). This survey was based on a sample of 85 design consultancies. The second survey was conducted in 2006, and the results are reported in MacPherson and Vanchan (2007). This survey was based on a sample of 68 large U.S. producers of durable goods (those that engaged in producing goods that typically last for more than one year), giving a user-based perspective. Although we have no wish to reiterate the methodological or empirical details in this chapter, the two surveys provide a springboard for the discussion of industry themes that warrant further attention from an empirical standpoint.

Supply-side perspectives

Most of the design companies that responded to our 2005 survey were micro-businesses with fewer than ten employees, and most were less than 20 years old at the time of the survey. A majority served corporate clients in the manufacturing sector, notably on the basis of repeat business. Most firms were found to compete primarily on the basis of service quality, reputation, or image – with few respondents indicating a significant role for service cost (i.e. price competition). The main sectors served included aerospace, automotive products, household goods, apparel/textiles, and machinery. With the exception of firms serving fashion-based markets (e.g. women's clothing), most firms were established by individuals with at least five years of prior experience working inside the manufacturing sector or for other design consultancies/studios. Many of these firms now serve their former employers, as well as other clients in related sectors.

The survey uncovered three distinct groups of firms in terms of service diversity. The first group (Tier-I) consisted of 16 firms that offer a full range of new product development (NPD) services, including design research, product or component design, and marketing services. These NPD vendors belong to an elite group in terms of the geographic range of their sales territories. The second group (Tier-II) consisted of 35 companies that offer design and marketing services, whereas the third group (Tier-III) comprised 34 companies that offer design services only. No statistically significant differences were found between TPPs (Total Package Providers) and other industrial design firms in terms of their age, employment size, or occupational structure (P values > 0.05; for details see Vanchan 2007).

Members of the Tier-I group were found to outperform their Tier-II and Tier-III counterparts in terms of several dimensions, including export-intensity, investment in worker training, research spending, recent rates of sales growth, and employment creation. For example, Tier-I firms exhibited faster growth rates. They were more likely to invest in worker training and R&D, and to create more new jobs than their lower-tier counterparts (for statistical details, see Vanchan and MacPherson 2008). The critical difference between Tier-I companies and their less diverse counterparts is that the former are qualified to undertake design research and/or design tradeoff analysis (DTA). Under DTA contracts, for example, clients expect their vendors to profile statistically the commercial merits of potentially competing design objectives such as product performance, durability, ease of maintenance, buildability, or market appeal. From our sample, it would appear that only Tier-I companies have the competence to do this. This is due, in large part, to the fact that Tier-I companies employ specialists that have received formal academic training in design research.

From a service dynamics perspective, design research and DTA represent the fastest growing areas of specialization within the U.S. industrial design sector. An important implication is that analytic capability is becoming a competitive requirement, and that good design skills alone may not be sufficient to sustain companies that operate outside the various fashion-based domains (e.g. jewellery or apparel). Our survey results also indicated that design companies are increasingly being asked to provide services that pertain to industrial or commercial futures (e.g. technological forecasting and trend analysis).

Follow-up telephone interviews with 29 design companies yielded insights that we had not anticipated during the early stages of the inquiry. These interviews included all 16 of our Tier-I companies, along with 6 from Tier-II and 7 from Tier-III. Although we cannot report the full range of results here, some of our key findings warrant brief mention. First, the design industry appears to be split into multiple groups when it comes to competitive strategy and organizational structure. Among Tier-III producers, for example, aesthetic or artistic goals appear to drive the business development process. Four of our 7 interviewees from this group design women's clothing or fashion accessories, where the client typically seeks branding rights to advertise a particular name. These types of firms do not need to conduct design research or DTA. Instead, they operate on the basis of image – and rarely create new jobs. In contrast, 11 of our 16 Tier-I companies operate outside the iconic class. These firms provide blueprints for items that are less image-dependent – though it certainly does not hurt if such items look good. In between these two classes, the Tier-II companies straddle a middle-ground – where product functionality must be balanced alongside aesthetics or ergonomics (e.g. toothbrushes).

Interestingly, our follow-up interviews underscored a series of important points that we missed in our earlier publications. For example, a serious problem with any kind of statistical analysis of firm-level performance in the design industry is that standard metrics of business success may be inappropriate (e.g. sales growth or job-creation). For example, one firm in our sample is owned and managed by a well-known female designer that creates inimitable blueprints for women's shoes. She has no wish to grow her company beyond its existing size (one designer, one secretary), she makes loads of money, she has never been concerned about import competition, and she only works about 15 hours per week. Is she less successful than our fastest-growing Tier-I company, that designs electronic sub-systems for Airbus, Boeing, or Lockheed-Martin?

This said, 14 of the 16 Tier-I companies indicated that their final outputs are typically produced with specialized help from external partners (i.e. network-based collaboration). In contrast, only 4 of the 13 non-Tier-I companies operate with such networks. This might explain why so many of the smaller Tier-I companies are able to offer a full array of NPD services (e.g. prototype development, testing, research, design, and marketing). Our interviews also revealed that risk-sharing contracts with clients are more common among Tier-I design companies than among their less diverse counterparts. Part of the reason for this contrast is that Tier-I companies are more likely to participate in projects that require radical innovation in untested fields (e.g. the use of composite materials rather than metals). 12 of our 16 Tier-I companies noted that risk-sharing agreements are required by clients, and that this puts a strain on cash-flow because payment streams are dependent on the future market performance of final products. Although such agreements create strong incentives to deliver superior designs or other NPD inputs, most design firms would prefer to bill by the hour – which is a common practice outside the Tier-I domain. Even so, most of our Tier-I interviewees indicated that risk-sharing contracts can deliver major dividends for successful projects (i.e. revenue sharing).

Finally, most of our interviewees expressed modest optimism regarding their future commercial prospects – regardless of service diversity levels. Some of the more common reasons for this optimism included: (1) the belief that U.S. durable goods producers will continue to outsource specialized design tasks in an effort to boost efficiency; (2) the perception that design excellence is becoming a major priority among clients, especially in light of rising import competition from cheaper producers; and (3) the belief that there is a growing perception among both household and industrial consumers that product quality matters more than price.

Demand-side perspectives

In an effort to explore the demand side of the design equation, we surveyed the top 100 U.S. manufacturing firm in the summer of 2006 (for methodological and statistical details, see MacPherson and Vanchan 2007). We achieved a 68 per cent response rate, which is respectable. Of our 68 participants, 17 were found to operate almost entirely on the basis of in-house design activity (these were mostly producers of military items). Some of the key findings of the 2006 survey included the following. First, externalized design activity among the 51 outsourcers moved from 21.8 per cent in 1995 to 36.6 per cent in 2005 (an increase of 67.8 per cent over ten years). By 2005, the survey firms allocated more than one third of their total design budgets to external vendors. Second, independent consultants accounted for 40 per cent of these external expenditures in 2005 (compared to 24 per cent in 1995). Third, offshoring of design services increased from 15 per cent to 35 per cent over the study period (an increase of 133 per cent). The main foreign sources included the UK (29 per cent), Italy (20 per cent), France (15 per cent), and Germany (14 per cent). Fourth, 39 of the 51 outsourcers organized their external design contracts on a risk-sharing basis in 2005. And, fifth, outsourcers were found to outperform non-outsourcers in terms of several innovation metrics (i.e. rates of successful product development).

In terms of motives, two interlinked factors emerged as being especially important. First, around 70 per cent of our outsourcers indicated that they externalize part of their design requirements in an effort to focus in-house personnel on areas of core or strategic competence. Second, over 60 per cent indicated that design outsourcing is practiced to access entirely new types of services – or superior services. In both instances, however, the key point is that outsourcers typically do not want to replicate externalized activity via broader in-house effort. Instead, the goal is usually to capture missing ingredients in terms of fine details.

But these generalizations need to be tempered by caveats that pertain to sector membership. In the case of one firm, for example, design work is externalized to capture the names of specific fashion designers (i.e. an iconic approach). In other cases, design is externalized in the hope that creative gurus will come up with something radically new (Apple's iPod is a much publicized example). In still other cases, the goal is to cut costs in light of the superior efficiency of external vendors (Hewlett-Packard is known to do this on a regular basis). On balance, however, it is fair to suggest that the prime contractor has generally invested a good deal of time in initial concept design or product definition prior to externalizing specific tasks.

There is at least some evidence in the recent literature that outsourcing can stifle innovation, and that strategic functions such as industrial design ought to be kept in-house as much as possible (see Dankbar 2007). Evidence from our 2006 survey suggests that negative innovation effects are not in place, and that most of the nation's most successful durable goods producers outsource specialized design work because they are innovative. In comparing the innovation records of outsourcers versus non-outsourcers, our goal was not to demonstrate the superiority of one strategy over another. Instead, our goal was to test for an innovation dampening effect. No such effect could be detected. Most of our outsourcers conduct extensive design and R&D activity in-house, and subcontract mainly during the early phases of a new product launch.

Summary and conclusions

Our two surveys mesh closely in terms of several broad themes that are starting to attract attention in the literature on industrial competitiveness. On the demand side, for example, imports of design services account for over 30 per cent of total external consumption today – compared to only 15 per cent ten years ago. On the supply side, it was found that close to 40 per cent of surveyed firms indicated that new competitors are mainly foreign. Interestingly, Tier-I suppliers were found to be less susceptible to import pressures than their less diverse counterparts. Many of these Tier-I companies earn over 10 per cent of their current revenues from export markets, with most expecting to see significant foreign sales expansion over the near future. Our first empirical commonality, then, is that international trade in design services is growing.

The main target markets for U.S. design service exports are virtually identical to the primary import sources by nationality. Specifically, advanced industrial economies such as the UK and Italy are major export destinations and import sources (i.e. intra-industry trade). At this point in time, newly industrializing countries such as China and India do not figure prominently in any of this trade. But will this always be the case? We suspect not, but have no data from which to generate a timeline.

Our empirical results also mesh with earlier streams of empirical research on the role of advanced producer services in regional economic development (see Beyers 2005). As reported by Beyers and Lindahl (1996), cost-driven externalization has not been the only driver of the outsourcing thrust. When it comes to the creation of external partnerships for knowledge acquisition, outsourcing at the high end of the sophistication scale has been powered by in-house technical deficiencies among manufacturers. However, cost-containment is at least an implicit factor in all this, in that users of external design services would surely find it prohibitively expensive to generate internally such inputs for every product launch. Another factor worth considering is that firms from both of our samples stressed the centrality of product quality as a design objective, a common theme being that high-cost producers can compete with cheaper producers by making better products.

A final point worth underscoring is that most U.S. design companies are tiny, and that some of the most visible and sought-after consulting units employ only one person. Clearly, these people must be very good at what they do. But how did they get to be so good? An interesting direction for future research might be to explore qualitatively the origins of these micro-businesses, especially with regard to the educational or employment histories of company founders. Vinodrai (2006) has done a good job on this topic in a Canadian context, but comparable research in the U.S. is sparse.

References

Beyers, W.B. 2005. 'Services and the changing economic base of regions in the United States', *The Service Industries Journal* 25:4, 461–76.

Beyers, W.B. and Lindahl, D.P. 1996. 'Explaining the demand for producer services: Is cost-driven externalization the major factor?', *Paper in Regional Science* 75, 351–74.

Bureau of Economic Analysis 2008. Industry Economic Accounts. (http://www.bea.gov).

Bryson, J.R. and Rusten, G. 2005. 'Spatial divisions of expertise: Knowledge intensive business service firms and regional development in Norway', *The Service Industries Journal* 25:8, 959–77.

Daniels, P. and Bryson, J.R. 2005. 'Sustaining business and professional services in a second city region', *The Service Industries Journal* 25:4, 505–24.

Dankbaar, B. 2007. 'Global outsourcing and innovation: The consequences of losing both organizational and geographic proximity', *European Planning Studies* 15, 53–61.

Florida, R. 2002. 'The economic geography of talent'. *Annals of the Association of American Geographers* 92, 743–55.

MacPherson, A. and Vanchan, V. 2007. 'The outsourcing of industrial design by large U.S. manufacturing companies: An exploratory study'. *North American Regional Science Association Conference (Savannah, Georgia, 9 November 2007)*.

U.S. Census Bureau 2007. 'County Business Patterns'. (http://censtats.census.gov/cbpnaic/cbpnaic.shtml).

U.S. Census Bureau 2007. 'Economic Census'. (http://www.census.gov).

U.S. Census Bureau 2007. 'American FactFinder'. (http://factfinder.census.gov).

Vanchan, V. 2007. 'Communication and relationships between industrial design companies and their customers,' *The Industrial Geographer* 4:2, 28–46.

Vanchan, V. and MacPherson, A. 2008. 'The competitive characteristics of US firms in the industrial design sector: Empirical evidence from a national survey', *Competition & Change* 12:3, 262–80.

Vinodrai, T. 2006. 'Reproducing Toronto's design ecology: Career paths, intermediaries, and local labor markets', *Economic Geography* 82, 237–64.

Wood, P. 2005. 'A service-informed approach to regional innovation – or adaptation?', *The Service Industries Journal* 25:4, 429–45.

5
The Geography of Producing and Marketing Design for Montreal Fashion: Exploring the Role of Cultural Intermediaries

Norma Rantisi

Introduction

The significance of design as a source of value-added for manufacturing goods is gaining wide currency in an era of globalization. Design offers a means by which firms in a given nation can differentiate their products and remain viable as markets become more open, competitive and volatile. However, as an activity that simultaneously entails aesthetic, technical and commercial considerations, it can not rely on an established set of practices. Rather, design will often involve short-lived cycles that require continual flows of information, with innovation in this context implying a command of current trends. Moreover, these cycles assume different temporalities for different sectors, with clothes having relatively shorter cycles in comparison to cars, for instance. With the rise of new communication technologies, information flows through the clothing supply chain (from textile supplier to distributor) have ushered in a new paradigm of 'fast fashion', which shortens the product cycle even further. Thus, the basis of competitiveness in this context is an ability continually to tap into market trends and to embody such trends within the design and marketing of one's products.

Indeed, nowhere is the up-to-date and information-intensive nature of design more apparent than in the field of fashion. As the paradigm of planned obsolescence, fashion is subject to anywhere from two to six fashion seasons (or 'cycles') per year. Thus, while conventionally viewed as a 'low-technology' sector, the design process within fashion demands extensive information as well as continuous engagement with – and coverage by – the media for the realization of a cycle from production through consumption. This is even more challenging in a context of globalization, as information can be collected from an ever-widening range of sources, such that the challenge is not one of merely acquiring information but of acquiring and disseminating the right kind.

This chapter provides an analysis of how independent fashion designers and high-end apparel manufacturers in Montreal, Canada contend with this challenge. Independent designers here refers to firms in which the designer is both owner and creative director and high-end apparel manufacturers refers to firms with in-house design capability, where the designer or design team provide creative direction and are often employees. In the discussion of the Montreal case, the term 'apparel producers' is used when referring to both groups. By drawing on over 40 interviews with key industry actors (including 28 apparel producers) conducted over the years 2005 to 2007, this study considers how designers and manufacturers obtain – and transmit – key market information, highlighting the role that cultural intermediaries play in facilitating this process. More specifically, the focus of the analysis is on intermediaries working in retail-oriented activities. While there is a broad range of actors in the industry who contribute to the design process (e.g. McRobbie 1998 and Rantisi 2004), a focus on the retail side corresponds to a recognition in popular and academic literature of an increase in the buying power of retailers since the 1990s (see Crewe and Davenport 1992; Wrigley and Lowe 1996) and the considerable influence they exert in shaping consumer markets.

An analysis of these dynamics in Montreal is particularly instructive, as Montreal has long held a status as the centre of apparel in Canada. Moreover, the industry is increasingly integrated within a global and especially a continental marketplace as local markets become increasingly saturated with imports, and apparel producers are grappling with the challenge of how to adapt to new markets and sensibilities. The Montreal case can provide fertile ground for analysing whether their ties to 'non-local' retail-oriented actors are markedly different from those of local ones and how, if at all, these linkages impact their ability to compete within the global marketplace.

This chapter begins with a review of the recent literature on the role that geography plays in the innovation practices of firms, with a particular focus on the design and marketing practices of cultural products industries. Then, a brief description of the global and local challenges that Montreal fashion firms face in a period of globalization is provided. The chapter proceeds to present an analysis of the nature of encounters between designers and buyers within the local market, followed by an analysis of the nature of encounters between designers and retail-oriented actors within 'non-local' markets. Lessons from the Montreal case for both theory and policy are taken up in the concluding section.

Design as a contextual process

Design is a creative activity that shapes the form, content and symbolic meanings of objects. As a practice, it embodies both objective and subjective dimensions (Bryson et al. 2004). The objective dimensions correspond to commercial and technical attributes aimed at satisfying the needs of clients, such as cost, materials and timing. The subjective dimensions, on the other hand, correspond to cultural attributes, denoting fashion trends, tastes and ethics (Bryson et al. 2004: 4). A blending of such commercial and cultural imperatives, however, entails great risk and uncertainty, particularly as aesthetic valuations are also critical to the process of creative production, and business criteria defined on the basis of profit, cost-benefit analysis or market research prove inadequate in accommodating such valuations (Banks et al. 2000; Santagata 2004).

Aspers (2006), a sociologist, captures the complexity of creative activities such as design by defining creativity as the production of 'novel ideas that are seen as useful' (478). For Aspers (2006), knowledge of what is useful is, by definition, 'contextual knowledge'. It is knowledge that is time- and place-specific and shaped by the broader set of economic and artistic relations in which an individual creator is implicated. Moreover, a particularly significant dimension of this contextuality is the final consumer market, which ultimately determines a product's value. While previous studies have noted the significance of consumers in validating (or not) creative work (e.g., Becker 1984), Fine and Leopold (1993) have noted that the realms of consumption and production have been for the most part treated as distinct spheres. Yet recent studies, such as those by Aspers (2006) and Entwistle (2006), contend that actors at the consumption end, who are closest to the consumer market and who have their pulse on the latest trends, can shape the production, as well as the reception, of ideas and meanings.

One scholar who has been instrumental in theorizing the links between production and consumption within the creative process is the sociologist Paul Hirsch. Hirsch (1972, 2000) was one of the first to present cultural products as the outcome of a chain of interrelated activities, ranging from conception to manufacture to distribution. More specifically, Hirsch (1972, 2000) highlights the role of cultural intermediaries (for example, marketing and distribution organizations at the consumption end) as central actors in the production of a cultural good. These intermediaries serve as gatekeepers who identify and promote particular categories of goods (Hirsch 1972). Through their gatekeeping function, they not only mediate tastes for the consumer but qualify and requalify a product's image and symbolic value, and offer creative stimulus for future rounds of design (Hirsch 1972, 2000; Rantisi 2004). Thus, far from being a linear process, knowledge of what to produce arises from a set of non-linear and complex relations, whereby creators, together with intermediaries, constitute a cultural-industrial system. And within the system, it is the nature and form of encounters between constituent actors that prove critical for 'contextualizing' creative activities such as design (see Entwistle 2006; Aspers and Skov 2006; Leslie and Rantisi 2006).

Does 'context' imply 'place'?

In viewing 'contextual' creative knowledge as a 'process' rather than an 'object' that can be readily transferred, Aspers (2006) posits that interpretation and understanding are key components of such knowledge. Yet, for economic geographers, the question remains as to whether or not such understanding can emerge over long distances. Scholars such as Cooke and Morgan (1998), Malmberg and Maskell (2002), Gertler (2003) and Storper and Venables (2004) emphasize the significance of proximity and place, suggesting that frequent face-to-face interaction is essential for building trust and a common cognitive framework between actors. This common cognitive ground can in turn facilitate a cooperative exchange of information and experiences. According to Storper and Venables (2004), proximity is critical for creating local 'buzz' – that is, conditions that give rise to knowledge flows based on purposeful exchange, the close monitoring of competitors or even chance encounters. Thus, the locale is viewed as significant for formal as well as informal, diffuse forms of exchange.

With respect to cultural industries, more specifically, a physical concentration of inter-related actors is particularly significant since, as noted above, they are engaged in creative interpretation. Asheim et al. (2007) contend that knowledge within these industries is of a symbolic form and is incorporated and transmitted in images, designs, artefacts, sounds and narratives. Consequently, it relies on a 'deep understanding' of the habits, norms and everyday culture of specific social groupings;

industry actors must be culturally 'embedded' and sensitive to idiosyncratic, often taken-for-granted qualities. Asheim et al. (2007) further suggest that the acquisition of such creative and interpretive skills can come from 'know how', that is, experience gained through practice rather than formal training, and from 'know who', that is, knowledge based on gossip or trends (or 'buzz') about potential collaborators with complementary competencies.

Over time, the physical co-presence of cultural actors can also give rise to other benefits. Scott (1996, 1999) and Molotch (1996) have illustrated that in addition to serving as a site of social reproduction in which cultural competencies are generated, an agglomeration of actors in a creative field contributes to the development of a distinctive set of signs and images that become tied up with particular locales. These cultural associations subsequently provide cachet and authenticity to cultural products and serve as inputs into the design and marketing of future collections. In this way, localization facilitates the collective production and appropriation of place-based aesthetics as well as information exchange.

A role for the 'global'

While literature on the significance of spatial proximity for the process of cultural production has gained credence in academic and policy circles, a contending perspective has emerged that highlights the importance of non-local linkages, particularly in a context of globalization. One strand of scholars within this competing perspective contends that when firms are embedded in localized industry networks, there is a risk that such firms will become increasingly insular and that this insularity will inhibit their ability to be creative and think 'outside the box' (see Grabher 1993; Bathelt et al. 2004). To overcome this tendency, firms can benefit by forging non-local ties that link them up to new networks and by extension, new ideas and practices. Exposure to these new economic networks can, in turn, widen the pool of stimuli (or the 'selection environment') from which a firm can draw to create new products, or alternatively, provide new mediums through which they can capitalize on and compete with a local design idiom.

A second strand of scholars argues that at a time when firms are (and should be) engaging in trade, a privileging of the local may detract from an ability to respond to distant markets (Malmberg and Power 2005). With consumption increasingly viewed as an integral part of the process of cultural production, ties to actors at the consumption end takes on a new urgency for producers. Scholars such as Whitely (1991), for example, cite the value of being 'close' to consumers, even geographically distant ones, in order to forge ties and tap into their cultural sensibilities. More recently, economic geographers further contend that since local producers and distant consumers still operate within the same supply chain they are likely to share common understandings of industry dynamics that could serve as a foundation for forging closer ties, what some dub 'relational' as opposed to 'spatial' proximity (see Amin and Cohendet 2000 and Allen 2000).

Such an argument is particularly relevant for the case of design, raising a number of issues. For example, how about when the designed object is a sensual one, one that is imbued with material attributes which are tied up with a symbolic image that must be seen or touched? Can the vision (and the message) of sensual designs be jointly constructed, communicated and promoted over long distances? What processes can ensure that designs are 'localized' in external markets? The analysis here seeks to address this question by comparing the experiences of Montreal designers within the local market with their recent efforts to promote products in the U.S. market. As noted earlier, the analysis of the Montreal case will be centred on the buyer-supplier interface, an interface that is at times direct and at times mediated by other actors within the industrial system.

Situating Montreal fashion within the North American marketplace

Montreal remains the primary urban centre for apparel in Canada. While the province of Quebec accounts for approximately 55 per cent of national employment in the industry, Montreal accounts for over 85 per cent of that total. The city is home to major international companies such as Parasuco, Peerless and Hilary Radley, and now ranks third in terms of apparel production within North America, following Los Angeles and New York. The bi-annual Montreal Fashion Week, initiated in 2001, and the annual Fashion & Design Festival of Montreal serve to solidify its status as a national fashion centre. An overview of the key institutions that constitute the fashion industry can be in found in Table 5.1.

Despite the significant presence of the industry, as in most advanced capitalist economies, the apparel sector has been subject to extensive restructuring in the last several decades due to a number of parallel trends. A process of globalization has been spurred on by advances in transport and communications technology and intensified by a

Table 5.1 Largest ten retailers' share of total Canadian apparel market, retail dollar sales (January–December 2006)

Retailers	Market share (%)
Sears*	13.6
Wal-Mart*	8.4
The Bay*	5.9
Zellers*	5.5
Mark's W.W.	4.3
Winners*	2.9
Moores	1.9
Reitman's	1.8
The Gap (Total)*	1.8
Costco*	1.6
Total %	47.7

* U.S. retailers.
Source: Trendex North America 2006 Annual Retail Sales Information, http://www.trendexna.com/annual_retail_sales_information.htm

liberalization of trade barriers, leading to an increasing trend toward rationalization and outsourcing. Two critical moments in this broader trend include the North American Free Trade Agreement (NAFTA) in 1989 and the World Trade Organization's Agreement on Textiles and Clothing (ATC), 1995–2004, which called for a gradual phasing out of quotas on imports. A consequence of these developments has been a surge in imports (see Figure 5.1) and significant declines in employment. Moreover, a concentration of buying power among the major retailers, such that the top ten retailers now control over half of apparel sales (see Figure 5.2), means that apparel designers and manufacturers face heightened competition for fewer and fewer accounts. Thus, while domestic retail sales in Canada have increased over the last decade, a smaller percentage of this demand is being met by domestic producers.

One way that apparel producers have responded to recent developments is by focusing on exports as a means to extend market share, with the U.S. targeted as a primary destination due to its proximity and large domestic market. In the wake of NAFTA, this strategy has been relatively successful, as exports to the U.S. increased ten-fold during the period 1990–2000 (Wyman 2006). According to Statistics Canada, in 2006 exports constituted nearly 40 per cent of Canadian apparel shipments, and the U.S. accounts for 88 per cent of total exports. For Montreal,

100 Norma Rantisi

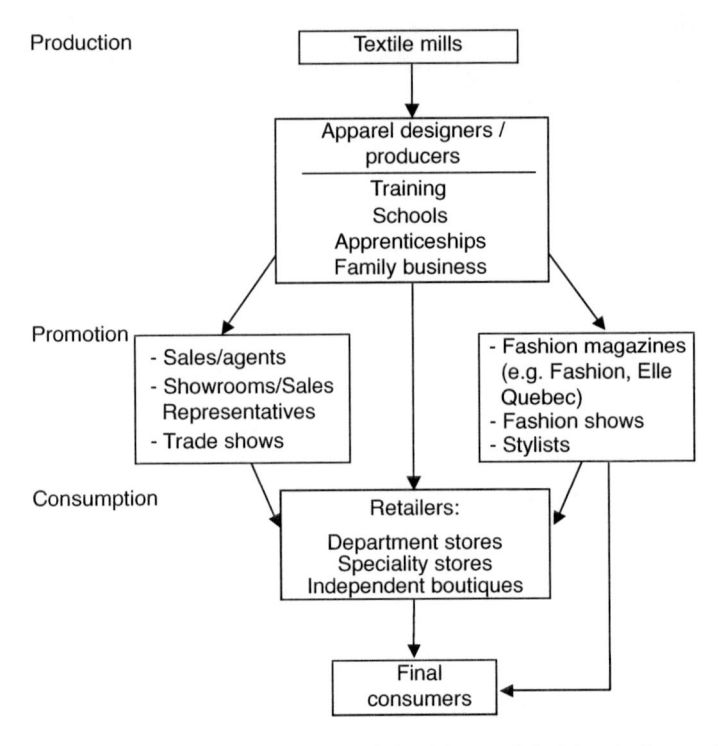

Figure 5.1 Overview of the structure of the Montreal fashion industry: from production to consumption

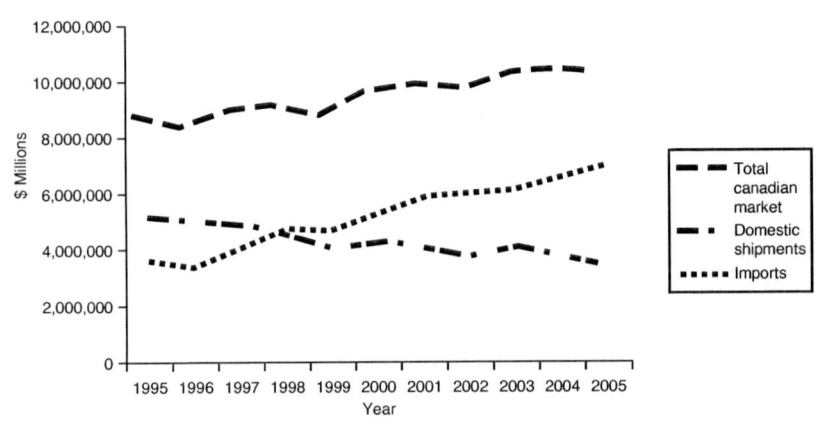

Figure 5.2 Canadian apparel market – domestic and imports (1995–2005)

Source: Apparel & Textiles Directorate, Service Industries and Consumer Products Branch, Industry Canada, and Statistics Canada data; manufacturing data are based on the North American Industry Classification System (NAICS) code and trade data are based on the Harmonized System (HS).

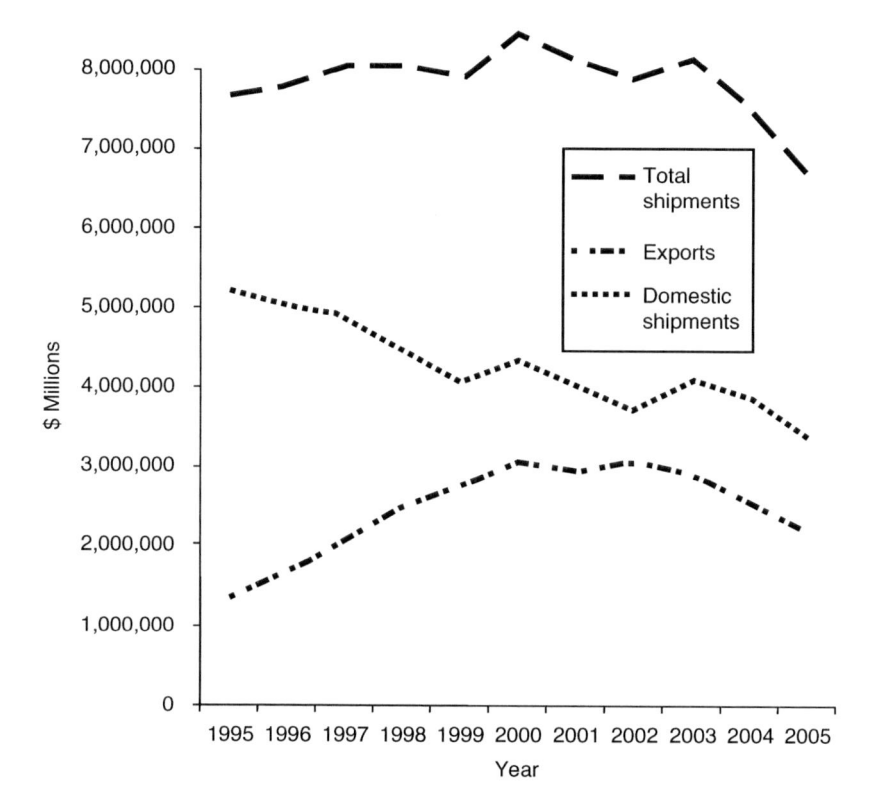

Figure 5.3 Canadian apparel market shipments (1995–2005)

Source: Apparel & Textiles Directorate, Service Industries and Consumer Products Branch, Industry Canada, and Statistics Canada data; manufacturing data are based on the North American Industry Classification System (NAICS) code and trade data are based on the Harmonized System (HS).

exports constitute roughly 20 per cent of the city's apparel shipments, with the U.S. accounting for 90 per cent. However, since 2002, a growth in imports to the U.S. from developing countries and a rise in the value of the Canadian dollar have led to a relative decline in exports (refer to Figure 5.3), presenting new challenges to securing and sustaining U.S. accounts. Competing on the basis of cost advantage is no longer an option.

And as criteria such as design and style become all the more significant, evaluations of current efforts by high-end manufacturers and designers to insert themselves into an increasingly global marketplace provide insight into the possible role that proximate ties to buyers – and by extension, to consumer markets – can play.

Montreal fashion's 'local' advantage

When examining designer-buyer links within the Montreal context, two distinctive features of the local apparel retail market become apparent. First, Montreal, in comparison to other metropolitan cities in Canada, has a large percentage of independent stores. While international retailers have been gaining market share more generally, in Canada (see Figure 5.2), local independent retailers in Montreal have been able to retain a presence due to the relatively affordable rents (interviews with independent retailer owners 2006, 2007). Second, and related, the cheap rents in Montreal also afford independent designers with the opportunity to open their own boutiques. Of the 24 apparel producers interviewed, nearly 60 per cent said that they owned their own retail shops. As illustrated in the next two sections, these features ensure that relations between suppliers and buyers (who are, most often, the store owners themselves) are direct or even overlapping. Such relations, in turn, condition how the meaning and value of designed objects are realized and (re)presented.

The role of independent stores in shaping the conception and reception of Montreal designs

For Montreal designers, the local independent retail scene offers a stepping stone towards building up a market. To initiate contact, designers will often make 'cold calls' or send out pamphlets or promotional packages to prospective clients. They may even go directly to the store to visit the owner (interviews 2005–2006). In an industry that relies heavily on personal networks, however, the most successful connections are made through word-of-mouth. Gossip among industry peers keeps buyers abreast of new or emerging talent and producers aware of potential accounts. One store owner, who is also the buyer, states that having the right network is key: 'Nobody knows everything...[so] you surround yourself with people that you believe in, people that you like their style, their taste and you share' (interview 2007). And one designer discussed the significance of the local 'fashion family' in contributing to his success (interview 2006). He talked in particular about the importance of building recognition and relations with his clients, journalists from newspapers and fashion magazines, buyers and stylists – those that provide visibility and promotion for his styles as well as general support and encouragement.

In some cases, the bi-annual Montreal Fashion Week, initiated in 2001, allows designers or high-end manufacturers to further build their visibility within the industry and to secure new accounts, particularly with the more established Canadian or Quebec-based specialty stores, such as Holt Renfrew and Simons. Of the apparel producers interviewed, nearly half have participated in Fashion Week, but less than one-fifth participate on a regular basis. One designer explained that it is cost-prohibitive to put a show together on a regular basis, with costs reaching roughly $5,000 per show (interview 2005). And the benefits of involvement do not materialize immediately, as it often requires several showings before one can establish their presence (interview with a designer 2005).

Regardless of the means by which relations with local buyers are established, once established, they tend to be frequent and close. The reason for this is two-fold. On the one hand, the designer can benefit greatly from the feedback they receive from the buyers. They can learn about what features of a product the end consumer likes or what elements can be adapted to better suit a store's clientele. Since most store owners are also buyers, they interact directly with both end consumers and with designers and thus, serve as critical conduits for market information. But information does not necessarily flow in one direction, from buyer to designer. Having close and frequent ties to buyers and other sales people enables a designer to convey important information about their product, about how it should be worn and displayed. One designer describes it as a process of translation: 'us, we do creative work that is a bit more artistic, it needs more explanation, it needs a person to tell [the retailer] how to do things' (interview 2006).

For their part, the buyers also benefit from direct communication with designers and from being able to work with them to ensure the development of viable products. This is especially essential in the early years of production for a designer. As one retail buyer notes, 'young designers...they don't really know what works and what doesn't. They have a concept and they work on it as hard as they can. They put it in the store. We see it on people. We let them know what is working and what is not, and they have to modify from there, if they want to continue in that direction...we develop [them] as much as we can without changing or stifling their style' (interview 2004). The fates of independent stores are often interconnected with those of the designers that supply them, since they rely on local, exclusive designs as their key selling point and as a means to distinguish themselves from chain stores[1].

Fashion designers as boutique owners: Transcending the production-consumption divide

In cases where fashion designers are themselves the boutique owners, they assume the role of creator and cultural intermediary concurrently and their ability to operate on both sides of the production-consumption interface is cited as a primary motivation for moving into the retail field. As one designer suggests, the link with consumers becomes 'direct' rather than 'mediated' (interview 2006). And with respect to the development and marketing of their products, there are numerous advantages that arise from this link.

As mentioned above, one critical advantage is the feedback that designers receive from their clients. Nearly half of the designers interviewed who own their own boutique have their design studio space on the same premises, generally at the back of the store (see Figure 5.3). Consequently, they spend a lot of time on the shop floor and in direct communication with their clients. Even designers that have studio spaces in a separate location often visit their stores and are in regular communication with clients and sales people (interviews with designers 2005–2006). In these encounters, comments about the style or fit of a clothing item naturally come up, in the form of compliments or complaints. One designer suggests that information about their products can also be obtained in indirect ways by 'being there': 'people come and you can see them; you can see them in your clothes, so you will notice things, like the fit' (interview 2006). And one designer goes as far as to equate their retail space with a 'laboratory', citing the value in having 'a place to test your creativity, your product, your fit' (interview 2006).

Another major advantage to having a boutique is the ability to control the image and the display of the clothing. A number of designers complained that they had little control over the presentation of their clothes in other shops, including where they were placed, which pieces were grouped together and the broader setting in which they were displayed. Presentation is a critical determinant to how a product is perceived and valued and by having complete control over the space, apparel producers could create a look and atmosphere that complements their line. This was conveyed by one high-end manufacturer: 'Our architecture has the same spirit as the product. The music we use has the same spirit. The people we hire know what to say about the product' (interview 2005). Moreover, in their own shops, designers can showcase some of the creations that people may not buy but that are still part of their overall concept (interview 2006).

Five of the interviewees also noted that a lack of service in the stores to which they sell is another impetus for having their own store. One designer complained that in department stores, in particular, salespeople do not generally assist consumers in the selection of garments or provide them with specific information about how a garment should be worn (interview 2006). As salespeople on the shop floor are deemed critical agents in the marketing and promotion of a product to the end consumer, control of this site along the commodity chain is viewed as an important part of the broader process of managing the product's image.

Above and beyond an ability to work closely with independent boutiques buyers or to own their own boutiques, Montreal apparel producers benefit from two other distinctive features of the local apparel market: the support of local fashion magazines and consumer loyalty. There are three magazines that are cited as particularly supportive of local talent: Elle Quebec and Clin D'Oeil, which are oriented to the Francophone Canadian market, and Strut, which was formerly based in Montreal and oriented to the streetwear market. There is also one national magazine, *Fashion*, that has a special section dedicated to covering regional markets. The primary form of support that these magazines provide is free coverage (interviews 2005–2007). One designer maintains that, in contrast to most of the national magazines, 'a lot of magazines in Quebec don't just show Chanel and Prada...at least a percentage of the editorials are Quebec designers. I really think that is an important thing. It helps us out, but it helps them out too. We're making their magazine more interesting' (interview 2005). In the same way that stores seek exclusive products to set them apart from the competition, magazines also seek to distinguish themselves in the marketplace and coverage of local, independent designers provides a form of distinctiveness.

Local apparel producers and independent store owners also cite the following of local consumers as a primary strength of the Montreal industry. This was cited by over 70 per cent of those interviewed and in most cases, small scale apparel producers credit their survival to local consumers who eschew international brands and seek out exclusive products. Within the Canadian context, Montreal has a reputation for being more artistic and less brand-conscious or corporate in style, when compared to other centres such as Toronto. Most interviewees attribute this unique aesthetic to the city's European influence and sensibility, and the fact it represents a mixing of cultures. Others attribute it to the fact that Montreal is not as affluent as other centres: 'there is more talent and eccentricity here in Montreal because of the whole starving artist syndrome. It's almost because there is not enough money and they're not selling that artists are not mainstreaming their product' (interview with a local designer 2005).

'Fashioning design from a distance': The role of intermediaries in bridging the local-global divide

While there are many advantages that accrue to Montreal apparel producers who are embedded in the local market, there is also a significant drawback to staying 'local' – the limited market. All the apparel producers interviewed cited the limited market as the primary disadvantage of being based in Montreal. This trend is also captured in numbers; within Canada, Quebec accounts for only 25 per cent of consumer expenditures in the apparel sector (as compared with Ontario, which accounts for 40 per cent) (http://www.infomat.com/research/infre0000291.html, accessed June 2007). Moreover, total apparel retail sales in Canada are less than one-tenth those in the U.S. (http://www.trendexna.com/images/COTW/06-11-07/cotw_061107.htm). Consequently, exporting to the U.S. is viewed as a new competitive strategy for a number of apparel producers. This strategy, however, entails its own set of challenges. Since the U.S. buyers market is significantly larger than the Canadian one, apparel producers require assistance for navigating the market – that is, they need to forge ties with people 'on the ground'. Moreover, there are not only a significantly larger number of buyers in the U.S., but there are also a significantly larger number of producers with whom an apparel exporter must compete. The challenge for Montreal apparel producers is to differentiate themselves in the marketplace – a challenge that is compounded by the fact that U.S. consumers are highly brand-conscious, whereas in Canada, few apparel producers will invest in developing a 'brand' by marketing to the end-consumer. Of the apparel producers interviewed, over 60 per cent are exporting, but less than one-sixth of those exporting (approximately 16 per cent) are investing in marketing via fashion magazines or other end-consumer channels (e.g. billboards). To date, Montreal apparel producers have focused primarily on promoting their styles to retailer buyers. The analysis below discusses in more detail the practices that they have employed to establish – and sustain – a presence in the U.S. market.

Sales representatives

The first step to establishing a presence overseas is to secure a sales representative. In most cases, apparel producers can not afford to hire their own exclusive sales representative so they must work with an independent sales agent, who has their own showroom and generally represents several lines at the same time. Designers and manufacturers interviewed noted that this was the essential starting point because the sales agents based in the U.S. market already have contacts with local retail buyers (interviews 2005–2006). These agents also have their pulse on local market trends, as they are in regular contact with numerous apparel producers and buyers at any one point in time. As one U.S.-based sales representative suggests: 'if everyone is coming in and everyone is looking for linen and linen is increasing in importance in the market, then we will tell them [the Montreal office]' (interview 2006). Although often overlooked in analyses of cultural production systems, these agents also mediate the buyer's encounter with the product (and by extension, the producer) in the way they display and present garments (see Entwistle 2006 for an elaboration of this). Thus, they serve as a critical intermediary and a valuable channel for information that would otherwise be difficult to acquire and transmit.

Yet the challenge for most apparel producers lies in identifying an appropriate agent to represent their collection. Ideally, a producer would find someone who is familiar with their specific price point and product segment and who can attract suitable buyers, someone both qualified and committed. As one designer explained: 'the most important thing is to find the right person to represent your collection. It's hard to find someone that trusts and believes in your collection. You also want to put your collection into a showroom that has a nice reputation, next to a product at the same level as you are. That is very hard' (interview 2006). Since independent agents cater to multiple apparel producers at once, they may not have the time to adequately learn about, or promote, all of the lines they carry, and they may give preferential treatment to some accounts relative to others. One high-end manufacturer complained: 'when they [sales agents] get a multiplicity of lines, you get a multiplicity of hats that you're wearing and a multiplicity of bosses. If the boss screams loud, he gets more attention' (interview 2005). Moreover, it may be harder for agents to provide the nuanced feedback to a producer that comes with knowing a particular line well.

Trade shows

Another means by which fashion firms access the U.S. market is through trade shows. Within the last two decades, trade shows have become the major venue in the U.S. for meeting potential buyers (Rantisi 2002). One of the most popular shows in the U.S. is MAGIC, which is held in

Las Vegas twice a year and caters to a range of product types and price points. There has also been a proliferation in the number and types of shows, many of which are regional (e.g. shows focused on the Midwest or Southern markets) and some of which are oriented to particular segments of the industry (e.g. a high-end show, such as, The Coterie in New York City).

The principal advantage of presenting at a trade show is that an apparel producer gains exposure to a large range of buyers who are concentrated in one location and who may not otherwise visit Montreal just to see an individual apparel producer. One designer noted that this is particularly significant for higher-end producers, since the established, wholesale showroom district in Montreal does not cater to this end of the market (2006). Several apparel producers who have been exporting for over five years said that their participation in trade shows were critical for building an 'identity' in the U.S. marketplace, emphasizing the importance of attending on a regular basis and of investing in the design of the booth to project a particular image. One high-end manufacturer and active trade association member said that many Montreal producers have been reluctant to make these long-term investments. According to him 'people go to shows for the wrong reason. They all go to the show and it costs $10,000. [The mentality is] "well, I've got to sell $25,000 or $30,000 to break even". That is the thought process, instead of saying "I'm building my brand and I'm building my market by exposing and being in front of the people and having an opportunity to press the flesh"' (interview 2005). While most apparel producers may send representatives or agents to the shows on their behalf, several discussed the value in going themselves and meeting face-to face with the buyers. Through these encounters, apparel producers can acquire direct feedback from customers about the performance of their lines. They can also use the encounter as an opportunity to 'explain' their new lines to both potential and established customers.

Showrooms

While the initial encounter with a prospective buyer may occur at a trade show, if the buyer is interested in seeing and learning more, then a private meeting at a showroom follows this encounter[2]. In some cases, this meeting might occur at the apparel producer's own showroom in Montreal. However, due to time constraints on the part of most U.S. buyers, it generally occurs at a sales agent's showroom in the U.S. – another setting over which the apparel producer has little control.

Given the significant role that the showroom plays as a site for product display and forum for buyer-supplier exchange, the most effective means to ensure and regulate their visibility in the marketplace is for apparel producers to acquire their own showroom in the U.S. and in particular, in New York City, the country's largest market. The advantages to having an independent showroom, in which a salesperson working exclusively for the company could be based, are numerous. First, a regular presence in the local market helps to nurture relations with key clients as well as with competitors, both of whom are critical sources of information on local trends. With regards to this advantage, an apparel sales executive based in New York maintained that, by having a regular presence, 'you have somebody that is living, eating, breathing the country, the competition, being with the buyers. It is incredibly important; you can not do it from outside of the country that you intend to operate in' (interview 2006). Another advantage to having one's own showroom is that it could be designed according to the needs and preferences of the designer or manufacturer. An apparel producer can create an atmosphere that coheres with the aesthetic of the product or line that they are selling and contribute to a product's representation.

Beyond these advantages, the showroom could also help to 'brand' an apparel producer and to project a sensibility to the local market. In one case, a designer had a showroom based in the heart of the New York garment district with her name on the window, clearly visible to buyers from the outside. Due to having a central and public location, the designer stated that her label was often mistaken for a U.S. brand (interview 2006). Moreover, a mark on the physical landscape not only projects this image to retail buyers but also to potential end consumers passing through the highly frequented district.

By providing a stable and exclusive presence, the showroom is integral to enhancing the direct and indirect encounters between buyer and supplier. It not only serves as the site in which such encounters occur but also as a medium through which information about a product can be transmitted, and for 'non-local' apparel producers in particular, it can give rise to many of the benefits that are typically associated with 'being there'. Yet, these advantages notwithstanding, to date, few have benefited from this strategy. Of the Montreal apparel producers interviewed, only five said that they had their own showrooms in the U.S. For most, it is too costly, given the high rents in major apparel centres such as New York and Chicago; therefore, working with independent sales representatives in the U.S. is viewed as the next best option.

'Montreal collections': A publicly-sponsored networking and branding program

In addition to private initiatives to permeate the U.S. market, there is a public program entitled Montreal Collections, which was set up by the provincial Ministry of Economic Development, Innovation and Export in the mid-1990s to promote Quebec designers in the U.S.[3] It is managed by a provincial office based in Montreal, but operates through 'government desks' (i.e. overseas officials) in six of the major apparel centres in the U.S.: New York City, Chicago, Atlanta, Las Vegas, Miami and Boston. The primary motivation for initiating the program was the view that, despite the appeal of the U.S. as a destination for Quebec apparel, Quebec – and even Canada – had no real meaning for U.S. buyers and sales agents (interview with the director of Montreal Collections, 2005). Thus, the government has sought to develop greater visibility for apparel firms through a set of initiatives.

A key initiative of the program is to have overseas government officials identify local sales representatives for potential apparel exporters. As mentioned above, this is a critical starting point for apparel producers and yet a challenge exists for most producers to identify appropriate sales representatives. Since overseas government officials are permanently based in the U.S., they are more easily able to develop ties to sales representatives and retail buyers in their respective markets and to build a contact database by visiting showrooms or by attending trade shows on a regular basis (interview with two overseas officials, 2006). These officials also promote networking between buyers and suppliers by providing apparel producers with temporary meeting and showroom space when they are visiting. Over time, as the overseas government officials become more acquainted with the local markets, they not only fulfil a networking function, but serve as an important source on market trends. They are able to provide apparel producers with the feedback that they receive from sales representatives or buyers. This was mentioned by one official, who stated that part of his job was to 'advise them as to what to do in order not to make mistakes, not to hurt their position in the market... general counselling as to what to do in order to succeed here' (interview with an overseas government official, 2006). Specific examples included suggestions about altering the sizing of products or altering the colours for different regional markets, and adapting how they communicated with U.S. buyers.

A second initiative of the program is a marketing campaign, whereby apparel producers are promoted under the common banner of 'Montreal

Collections'. Modelled after a similar Italian initiative, the campaign is based on the view that, since Montreal lacks any well-known brands, firms could benefit from collective marketing by being grouped with other firms hailing from the same locale. Montreal – as opposed to the nation or province – was selected as the trademark because fashion has historically been an urban phenomenon and because Montreal is positively received in the U.S. (interview with director of Montreal Collections, 2005). At present, the campaign is supported through a website that contains information about Montreal-based apparel exporters and through advertisements in U.S. trade magazines. The sponsor and promotion of Montreal apparel producers at U.S. trade shows is another critical part of the campaign. The program helps apparel producers identify the appropriate trade shows to attend and hosts publicity events at the shows. One such event is a fashion show dedicated exclusively to presenting the lines of Montreal Collections members. Fashion shows have been hosted at trade shows in Atlanta and Chicago, with as many as 20 to 30 members presenting in a single show. Other activities that are sponsored at trade shows include the provision of quintessential Canadian and Quebec cultural products, such as Molson beer, Quebec cheeses and even a circus performance by the Montreal-based Cirque Eloize (interviews with officials involved in Montreal Collections, 2006). Such activities are significant in that the tie-in with other well-known cultural products gives greater prestige to Montreal fashion and can shape the signification of fashion designs. Moreover, the cross-promotional marketing helps to elevate Montreal's status as a creative centre, which reinforces the cachet of being associated with – and 'branded' through – the city.

While the benefits of a program such as Montreal Collections are hard to quantify, there is clearly an interest on the part of apparel producers to have some association. The organization currently has 300 members. Of the 19 apparel exporters interviewed for the study, 11 are listed the organization's website and 3 others stated that they benefited from trade show sponsorship and/or from the provision of temporary showroom and meeting space in the U.S. (interviews 2005–2006), with 1 designer explicitly attributing his ability to enter the U.S. market to the support that he received from the program (interview 2005). There are still some limitations, however, to the extent that the program can contribute to the development, signification and promotion of Montreal fashion. First, the program is primarily restricted to those who are 'export-ready', that is, those who already have a sales person or are already selling outside the Montreal market (e.g. to other parts of

Canada or overseas). Many of the independent designers interviewed for the study had still not penetrated any markets outside Montreal; they were the apparel producers most in need of support in that they had little knowledge (or resources) about how to design and market for other contexts, and yet, they were the least eligible. Second, while collective marketing can raise the visibility of all firms involved, individual firms can not rely solely on this form of marketing. The director of the program and one overseas official emphasized the need for apparel producers to complement these efforts with individual strategies, since firms must ultimately distinguish themselves in the marketplace, and collective campaigns threaten to dilute the individual identity of a firm if not complemented by firm-based ones (interviews 2005, 2006). In particular, the officials noted the need for apparel producers to more seriously consider the importance of targeting the end-consumer, and not just people in the trade, in their efforts to develop a 'brand' or identity in the U.S. market. In the long-term, this strategy is viewed as ensuring a market (or a 'following'), so that an apparel producer is not so vulnerable to a change in the retail buyers with whom they work or to the loss of an account. Furthermore, by conveying the image of their products directly to consumers, in an 'unmediated' form, apparel producers can acquire greater credibility with retail buyers and become less reliant on salespeople in the showrooms or in the stores for the merchandizing of the product.

Lessons of the Montreal case for policy and theory

This analysis of the case of Montreal apparel producers highlights a number of opportunities and challenges relating to the design and (re)presentation of fashion from a 'distance'. On the one hand, it suggests that face-to-face contact is important for building a rapport with buyers that can facilitate the acquisition of nuanced information about how products may need to be altered; such contact also facilitates the transmission of information about the particular design qualities of a product from supplier to buyer. In cases where face-to-face contact is not possible, on the other hand, the role of intermediaries is critical. While there is always the potential for a loss of critical knowledge when producers (and products) are removed from a particular context and operating at a distance from the 'consumer', 'relationally' close intermediaries – those with in-depth knowledge of and influence over the product – play an important role in managing how a product is negotiated and received. This suggests that even 'relational' proximity can serve as an alternative, albeit an imperfect one, for 'spatial' proximity in the process of knowledge acquisition and dissemination.

From a policy perspective, a linking up with 'relationally' close intermediaries in a distant market proves to be the biggest challenge, particularly for those apparel producers who do not have their own showroom or a dedicated sales representative who works exclusively for them. In such cases, apparel producers may need to devise alternative strategies for embedding themselves or their products in a foreign market. One such strategy may be to apply a cross-marketing initiative, similar to that instituted by the Montreal Collections program. This could entail a Montreal fashion designer sharing the costs of a showroom with other Montreal-based designers who work in different but complementary markets, for example, jewellery designers or other fashion designers in similar product areas but with different price points. This could enable them to have a dedicated salesperson and a space that reflects their shared aesthetics; alternatively they could divide up a showroom into distinct sections. Another strategy would be to forge more direct links with consumers through restricted cross-marketing in fashion magazines – for example, an outerwear designer advertising with a pant designer or a dress designer with a shoe designer. Again, this can be a way to socialize the costs and to exploit the potential synergies of related creative products in ways that still retain the distinct identities of the individual producers. Closer relations to a geographically distant market could also be forged by seeking out buyers for specialty or independent stores rather than large department stores. Several of the apparel producers have stated this intention, since the buyers for specialty/independent stores tend to have a greater interest in the quality and design of the product and are not merely oriented to cost reductions. Thus, they are more likely to provide nuanced, qualitative feedback on market trends rather than blunt or quantitative assessments. Moreover, as noted above, they have more at stake in learning about and properly conveying the meaning of a product.

At a more general level, both private and public actors can collectively support the development of a local design infrastructure, one that can provide designers with resources for marketing their designs and that can help to forge both local and non-local networks. A particularly significant element of such infrastructure is the design school, which can train designers in how to access and work with market information, and to better balance aesthetic creativity with commercial imperatives, that is, to overcome the production-consumption divide. Such programs would strengthen the viability of a local design community, which in turn would strengthen the local design 'identity' that becomes mentally associated with a city, contributing further to the market viability and reputation of a city's design products. In today's business context, where information is being acquired, processed and disseminated at ever faster rates and product life-cycles are becoming shorter, such initiatives can go a long way in supporting the development and promotion of globally competitive products.

Acknowledgements

The research for this study was supported by Le Fonds Québécois de la Recherche sur la Société et la Culture, Programme d'établissement du Nouveaux Chercheurs.

Notes

1. Several have received free coverage in the local press for supporting local design talent (interviews 2005, 2006).
2. In cases where a city hosts a 'market week' (that is a dedicated week when buyers visit en masse to see samples for a given season) in lieu of a trade show, the showroom may even serve as the site for an initial buyer-supplier encounter.
3. Part of the mandate of this program is also to promote Quebec designers in other parts of Canada; it has local offices in Toronto and Vancouver, but the emphasis of the program is on the U.S.

References

Allen, J. 2000. 'Power/economic knowledge: symbolic and spatial formations'. In Bryson, J.R., Daniels, P.W., Henry, N. and Pollard, J. (eds) *Knowledge, Space, Economy*, 15–33. Routledge, London.

Amin, A. and Cohendet, P. 2000. 'Organizational learning and governance through embedded practices', *Journal of Management and Governance* 4:12, 93–116.

Asheim, B., Coenen, L. and Vang, J. 2007. 'Face-to-face, buzz, and knowledge bases: Sociospatial implications for learning, innovation, and innovation policy'. *Environment and Planning C: Government and Policy* 25:5, 655–70.

Aspers, P. 2006. 'Contextual knowledge', *Current Sociology* 54:5, 745–63.

Aspers, P. and Skov, L. 2006. 'Encounters in the global fashion business: Afterword'. *Current Sociology* 54:5, 802–13.

Banks, M., Lovatt, A., O'Connor, J. and Raffo, C. 2000. 'Risk and trust in the cultural industries', *Geoforum* 31:4, 453–64.

Bathelt H., Malmberg A. and Maskell, P. 2004. 'Clusters and knowledge: Local buzz, global pipelines and the process of knowledge creation', *Progress in Human Geography* 28:1, 31–56.

Becker, H.S. 1984. *Art Worlds*. University of California Press, Berkeley, CA.
Bryson, J.R., Daniels, P.W. and Rusten, G. 2004. 'Design workshops of the world: The production and integration of industrial design expertise into the product development and manufacturing process in Norway and the United Kingdom', Working Paper No 53, *Institute for Research in Economics and Business Administration*, Bergen.
Cooke, P. and Morgan, K. 1998. *The Associational Economy*. Oxford University Press, Oxford.
Crewe, L. and Davenport, E. 1992. 'The puppet show: Changing buyer–supplier relations within clothing retailing', *Transactions of the Institute of British Geographers* 17:2, 183–97.
Entwistle, J. 2006. 'The cultural economy of fashion buying', *Current Sociology* 54:5, 707–24.
Fine, B. and Leopold, E. 1993. *The World of Consumption*. Routledge, London.
Gertler, M.S. 2003. 'Tacit knowledge and the economic geography of context, or the undefinable tacitness of being (there)', *Journal of Economic Geography* 3:1, 75–99.
Grabher, G. 1993. 'The weakness of strong ties: The lock-in of regional development in the Ruhr area'. In Grabher, G. (ed.) *The Embedded Firm: On the Socioeconomics of Industrial Networks*, 265–77. Routledge, London.
Hirsch, P. 1972. 'Processing fads and fashions: An organization-set analysis of cultural industry systems', *American Journal of Sociology* 77, 639–59.
Hirsch, P. 2000. 'Cultural industries revisited', *Organization Science* 11:3, 356–61.
Leslie D. and Rantisi, N.M. 2006. 'Governing the design economy in Montréal, Canada', *Urban Affairs Review* 41:3, 309–37.
Malmberg, A. and Maskell, P. 2002. 'The elusive concept of localization economies: Towards a knowledge-based theory of spatial clustering', *Environment and Planning A* 34:3, 429–49.
Malmberg, A. and Power, D. 2005. 'On the role of global demand in local innovation processes'. In Shapira, P. and Fuchs, G. (eds) *Rethinking Regional Innovation and Change*, 273–90. Springer, New York, NY.
McRobbie, A. 1998. *British fashion design: Rag trade or image industry?* Routledge, London.
Molotch, H. 1996. 'L.A. as design product'. In Scott, A.J. and Soja, E.W. (eds) *The City: Los Angeles and Urban Theory at the End of the Twentieth Century*, 225–75. University of California Press, Berkeley, CA.
Rantisi, N.M. 2002. 'The local innovation system as a source of variety: Openness and adaptability in New York City's garment district', *Regional Studies* 36:8, 587–602.
Rantisi, N.M. 2004. 'The designer in the city and the city in the designer'. In Power, D. and Scott, A.J. (eds) *Cultural Industries and the Production of Culture*, 157–90. Routledge, London.
Santagata, W. 2004. 'Creativity, fashion and market behavior'. In Power, D. and Scott, A.J. (eds) *Cultural Industries and the Production of Culture*, 75–90. Routledge, London.
Scott, A.J. 1996. 'The craft, fashion, and cultural products industries of Los Angeles: Competitive dynamics and policy dilemmas in a multi-sectoral image-producing complex', *Annals of the Association of American Geographers* 86:2, 306–23.

Scott, A.J. 1999. 'The cultural economy: Geography and the creative field', *Media, Culture and Society* 21, 807–17.
Storper, M. and Venables, A.J. 2004. 'Buzz: Face-to-face contact and the urban economy', *Journal of Economic Geography* 4:4, 351–70.
Whitley, R. 1991. *The Consumer-driven Company: Moving from Talk to Action*. Addison-Wesley, Reading, MA.
Wrigley, N. and Lowe, M. (eds) 1996. *Retailing, Consumption and Capital: Towards the New Retail Geography*. Longman, Harlow, Essex.
Wyman, D. (2006) 'Trade liberalization and the Canadian clothing market', *Canadian Economic Observer* 11-010-XIB, Statistics Canada, December 2006 edition.

6
Designed Here, Made There? Project-based Design Work in Toronto, Canada

Tara Vinodrai

Introduction

In February 2004 the front cover story of *The Economist* highlighted the shift of white-collar work to offshore locations as firms responded to competitive pressures and lowered their production costs in a globalized economy. In doing so, *The Economist* asked the provocative question: what will be left if professional services relocate to offshore locations? How will cities, regions and nations in advanced economies remain competitive? According to a growing number of observers of the contemporary economy, the answer is very simple: creativity.

Creativity is now understood as key to regional and national competitiveness and economic success in the contemporary economy. Concomitant with this increased emphasis on creativity has been the growing interest and visibility of design as a creative profession and practice. Design is seen to be assuming a strategic role in the competitiveness of firms, regions and nations (Lash and Urry 1994; Nussbaum 2004b). Nowhere has the successful use of design been more visible than in the ascent of Apple's iPod emblazoned with the simple phrase 'Designed in California, Made in China'. This makes a very powerful claim about the underlying geography of design activity, especially in relation to other aspects of globalized production networks.

Taken together, these observations are highly suggestive of a distinctive spatial division of labour whereby design activity remains in a handful of large, global cities in open, advanced economies while production takes place in low cost, 'offshore' locations. Yet, the case of design highlights the contradictions that emerge by assuming such a simple division of labour, since design can be at once a creative activity and a professional service (Rusten and Bryson 2007). One tension that may arise if design work takes place far from production is the need to balance proximity to clients and markets with proximity to production facilities. This chapter contributes to our understanding of these issues and their importance to the relationship between design and national competitiveness through an analysis of design work in Canada.

The analysis relies on evidence from a variety of primary and secondary sources. There is no single source of quantitative data on design activity in Canada. For this reason, employment data are drawn from the Canadian *2001 Census of Population* and the *Labour Force Survey, 1987–2004*, information on the number and size of establishments in the design industry are taken from the 1998 and 2005 *Canadian Business Patterns* and other details are derived from the 2002 *Survey of Specialized Design Services* and the 2003 *Survey of Innovation*. In addition to these secondary data sources, I draw upon evidence from an in-depth case study of design work in Toronto, including 60 in-depth interviews with practicing graphic and industrial designers, as well as interviews with representatives of professional associations, local design schools and government officials conducted in Toronto between March 2004 and February 2005. The sample of designers interviewed for this study was roughly divided equally between those working for firms and agencies in the design industry, those working for firms in other industries (e.g. manufacturing, other cultural industries, public sector and so on) and those working for themselves and balancing several contracts with clients. Designers worked on a range of projects, including everything from the design of new furniture, home products and small machinery parts to packaging, branding and displays. The interviews were semi-structured, lasting anywhere from one to three hours and were recorded and transcribed by the author. In some cases, interviews were conducted with multiple designers working at the same firm allowing for corroboration as to the nature of practices at the firm-level, as well as thoroughly assessing the extent of local and non-local collaborations and interactions. One main goal of the interview was to determine the nature of designers' career development and labour market mobility (see Vinodrai 2006). Relevant to this chapter, the interviews also probed into the nature and organization of design work, as well as the extent of and challenges associated with local and non-local interactions with clients and other project members. The analysis is further informed by attendance at a number of design-related events, as well as participation in a local study of the design capacity of Canada and its cities that included academic, industry and government partners (Gertler and Vinodrai 2004; Design Industry Advisory Committee 2004).

The chapter proceeds by framing the discussion in the context of recent claims about the importance of design to firm- and national-level competitiveness. A growing body of work demonstrates the importance of creative and cultural activities (including design) in the contemporary economy and documents their organizational and territorial dynamics, highlighting the urban, project-based nature of work requiring high levels of interaction. The literature on innovation and learning is informative in unpacking these territorial dynamics since scholars highlight the importance of proximity or 'being there' as being critical to applied problem solving in the production process, as well as to sharing tacit knowledge and ideas (Gertler 1995, 2003, 2004; Storper and Venables 2004; Bathelt et al. 2004). Following this brief discussion, I provide an overview of design activity in Canada, focusing on the underlying locational trends and employment distribution of design activity in both sectoral and occupational terms. The next section provides an analysis of design work in Toronto. I suggest that the nature of design work and labour markets, as well as the importance of proximity to clients and markets, reinforce the localization of design work. Yet, designers often need to seek out strategies to overcome the challenges that emerge from needing to balance proximity to clients and markets with proximity to production facilities. This is especially relevant since production has increasingly taken place in 'offshore' locations. The analysis confirms that the actual work of design in Canada remains concentrated in only a handful of urban places and this geography is reinforced through social and labour market dynamics. Overall, the analysis points to cities as critical sites of creative and design activity. Moreover, the discussion highlights the role of design in contributing to firm, regional and national competitiveness and prosperity, particularly in open, advanced economies.

Design, innovation and creativity in the contemporary urban economy

It has only been recently that design has enjoyed the spotlight amongst business leaders and policymakers. Citing the wild success of products such as Apple's iPod and other electronics made by companies such as Sony, LG and Samsung, recent literature on the 'business of design' claims that design can be used as a strategic management tool giving firms an edge in the global marketplace and that the use of design is imperative for the economic competitiveness of firms, regions and nations (Kotler and Rath 1984; Nussbaum 2004a, 2004b; Merritt and Lavelle 2005; Martin 2006).

Lash and Urry (1994: 15) claim that 'the design component comprises an increasing component of the value of goods', resulting in the centrality of the design process and the increasing 'design intensity' of products and services across the economy. Bryson et al. (2004), Power (2004), Rusten and Bryson (2007) and others have noted that design is a hybrid activity that is often central in the innovation process. Design and design strategy can be applied throughout the product development phase and production process in order to reduce production costs, as well as to improve both the functional and aesthetic qualities of a product. Furthermore, the use of design in the branding, packaging and marketing of products and (sometimes) their associated retail environments helps to create a distinctive product identity that differentiates firms' products and services in local and global markets with the goal of increasing revenues and sales. Design capability can be incorporated into the firm by 1) having their own in-house design department; 2) employing designers as part of multidisciplinary teams in various facets of their business (for example concurrent engineering, product development, marketing); 3) hiring freelance designers to work on specific projects; 4) purchasing the services of an outside design consultancy; or 5) using some combination of the above four options.

Even if the firm does employ designers and design strategy, this in itself does not guarantee success. Recent research suggests that firms that employ designers in-house *and* hire the services of a design consultancy are more likely to be successful (Core 77 2002). However, most of the evidence concerning the effective use of design as a tool for fuelling innovation and ultimately securing value-added for the firm has emerged from Europe and a handful of other developed and developing countries (New Zealand Institute of Economic Research 2003; Power 2004; Danish Design Centre 2003; Design Council 2004; New Zealand Design Taskforce 2003). Overall, these studies show that design is a critically important source of economic value, raising levels of profitability and productivity and contributing to national economic competitiveness and performance. In addition to these studies, a number of academic studies have illustrated that the consumer electronics or other technology-intensive industries are not alone in their use of design to secure their position in the global marketplace. Firms in more traditional industries such as furniture, textiles and apparel have also been able to re-invent themselves through the effective use of design (Lorenzen 1998; Rantisi 2002; Leslie and Reimer 2006).

This use of 'creative' resources such as design resonates with the well-established literature in the social sciences that recognizes that creative, symbolic and aesthetic content and inputs are critical in the production of goods and services in the contemporary economy (Lash and Urry 1994; Scott 2001; Florida 2002). Scholarly work in this area documents the industrial dynamics of a number of creative and cultural industries, including film and television, new media, fashion, publishing, music and advertising (see Scott 2001; Power and Scott 2004). These studies consistently demonstrate that creative and cultural industries are often highly innovative, yet innovation relies less on scientific discovery and R&D in the traditional sense and more on other inputs, including artistic and design inputs (Scott 2004). The literature is quite revealing as to the underlying territorial organization and dynamics that underpin these types of activities and a number of common themes emerge providing a stylized account of the geography of creative activity (see Reimer et al. 2008). First, creative and cultural activity tends to agglomerate in major urban centres (Pratt 1997; Scott 2001; Power 2002). This is seen as advantageous to firms since they can access deep pools of highly skilled, creative labour. Creative workers, on the other hand, benefit from being able to develop their careers while living in 'cool', diverse urban environments and neighbourhoods, which provide inspiration for their work (Florida 2002; Lloyd 2006). In turn, elements of place become embedded in their creative outputs (Molotch 2002; Rantisi 2004). Second, some creative industries are very susceptible to rapid shifts in consumer demands and changing styles, necessitating higher rates of innovation and easy access to information about changing tastes (Scott 2004). Third, scholars have suggested that this mode of production and innovation involves different organizational forms such as project-based work (Grabher 2002; Christopherson 2002). Fourth, creative work often involves high levels of self-employment and individual risk (Ekinsmyth 2002; Vinodrai 2006). Thus, the nature of work itself requires highly developed (local) social and knowledge networks to tune into the necessary background 'noise' about jobs and projects (Grabher 2004; Currid 2007). Moreover, a critical mass of people doing cutting edge work in the same and related fields generates and facilitates access to local 'buzz' about the latest trends and leading edge practices (Bathelt et al. 2004; Storper and Venables 2004). Overall, the agglomeration of creative activity in large urban centres provides opportunities for creative workers to access these local networks, thereby facilitating the easy transmission of knowledge about jobs and market trends through intensive face-to-face interaction. In short, the nature of creative work demands achieving proximity primarily through 'being there' (Gertler 1995; Storper and Venables 2004).

Yet, the extent to which proximity is necessary for learning, innovation, problem solving and knowledge exchange has remained the subject of intense debate (Maskell and Malmberg 1999; Amin and Cohendet 2004; Gertler 2004, 2008). Recent currents in the literature suggest that these key activities can span firm, regional and even national boundaries (Amin and Cohendet 2004). There are a number of mechanisms that firms can employ to overcome the challenges associated with learning at distance even within project-based environments (Grabher 2004). For example, Bathelt et al. (2004) propose that firms develop carefully constructed and maintained relationships (pipelines) with distant collaborators and partners to achieve learning across organizational and geographic boundaries (see Owen-Smith and Powell 2004). Other studies have highlighted the importance of communities of practice, Internet-based communication and international travel by managers and other highly skilled workers (Coe and Bunnell 2003; Amin and Cohendet 2004; Gertler 2004). In this way, a firm or project team can successfully overcome the problems associated with learning and problem solving at a distance.

Design work – as a creative, project-based endeavour – presents an interesting case since a potential need arises to balance proximity to clients and markets with proximity to production facilities often located in other national settings. Furthermore, despite the accumulated empirical evidence that design is an important input in the innovation process and critical to firm, regional and national competitiveness, there remain few studies that examine the territorial dynamics of design work (see Power 2004; Rusten and Bryson 2007; Reimer et al. 2008). Moreover, until recently, there have been few studies that have examined these dynamics within the Canadian context. A number of studies have started to fill this gap in our understanding of the contribution of design within the Canadian context (Gertler and Vinodrai 2004; Vinodrai 2005, 2006; Leslie and Reimer 2006; Leslie and Rantisi 2006; Rantisi and Leslie 2006). This chapter contributes to this growing literature by providing insight into the geography of design work in Canada and exploring the social and labour market dynamics that underpin this geography.

What is the geography of design work in Canada?

To understand the spatial distribution of design activity in Canada, I begin by examining the prevalence of design activity from both an industry and occupational perspective. This distinction is important, especially in the case of design. A singular focus on the design services industry misses an important segment of designers: designers employed in other industries. By taking an occupational approach (Feser 2003; Markusen 2004), it is possible to capture one way in which firms in established and emerging sectors (i.e.firms *outside* of the design industry) utilize design resources. Figure 6.1 directly compares employment in Canada's specialized design services industry to employment in design occupations over the period between 1987 and 2004[1]. It shows that the number of designers in Canada increased from 59,700 to 104,900 while employment in the Canadian design industry grew from 25,600 to 47,700. In other words, industries outside of the specialized design services industry are making use of design-related expertise by employing designers directly. In fact, only 43 per cent of all designers work within the specialized design industry; designers are employed in a wide range of traditional and emerging industries, including manufacturing (21.6 per cent), information and cultural industries (7.4 per cent) and wholesale and retail trade (8.6 per cent).

By indexing employment, it is possible to directly compare the growth of employment in design activity to the overall Canadian economy. Figure 6.2 shows the change of employment in Canadian design activity over the 18-year period between 1987 and 2004 (1987=100). Growth in design activity – measured by occupation and sector – has outpaced the growth of the overall Canadian economy. The growth and subsequent decline of employment in the early 1990s is indicative of a recessionary period that affected most sectors of the Canadian economy. However, it is clear that since the mid-1990s, there has been significant growth both in the number of people employed in the design services industry and in the number of designers, with employment in the specialized design services industry showing more volatility compared to steadier growth in the number of practicing designers.

In addition to examining employment change, we can look at the growth in the number and size of firms in the specialized design services industry. Table 6.1 shows the distribution of firms by size in 1998 and 2005. In 1998, there were almost 3600 firms in Canada's specialized design services industry. By 2005, there were almost 4500 establishments and the industry was growing at an average annual rate of growth that was more than four times the overall Canadian economy[2]. The industry is dominated by very small establishments, although the evidence suggests that average firm size increased over the period between 1998 and 2005. For example, in 1998, 68.2 per cent of firms

124 *Tara Vinodrai*

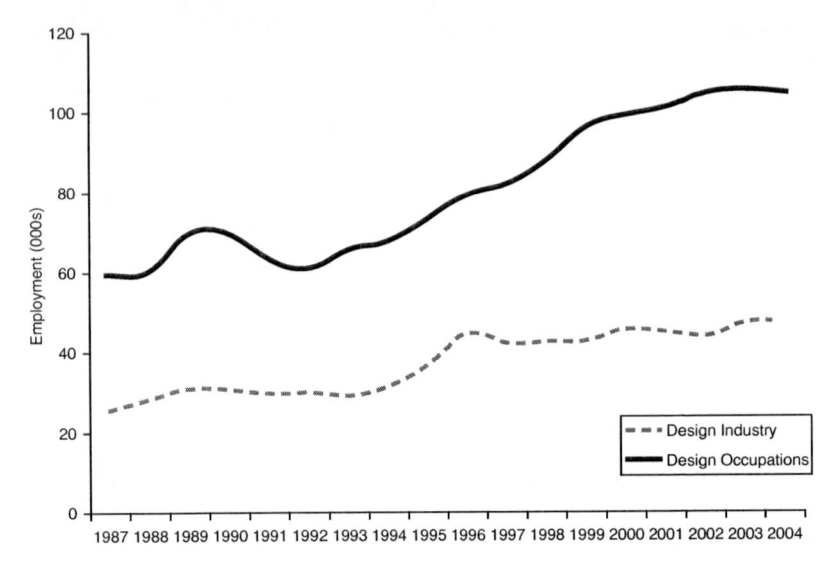

Figure 6.1 Employment in the Canadian design sector: Industry vs. occupation (1987–2004)

Source: Statistics Canada. Labour Force Survey, 1987–2004 [custom tabulations] author's calculations

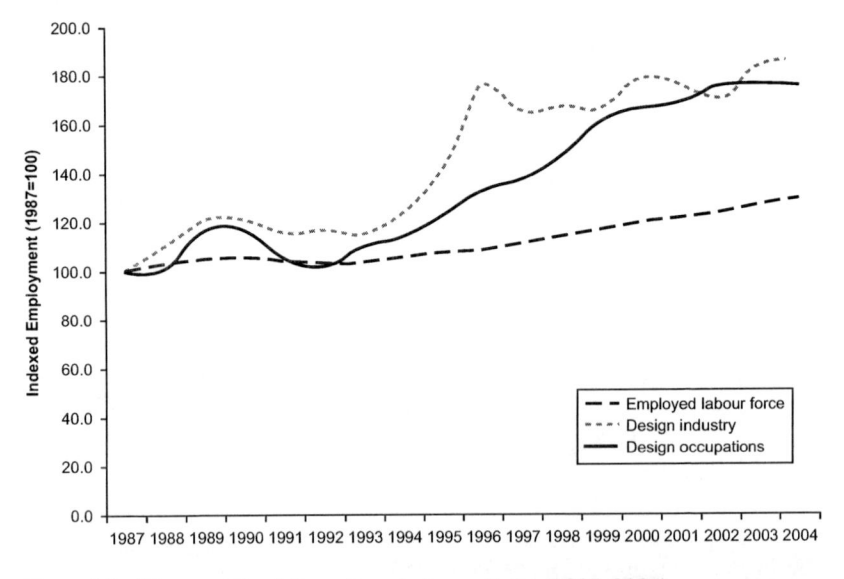

Figure 6.2 The growth of Canadian design activity (1987–2004)

Source: Statistics Canada. Labour Force Survey, 1987–2004 [custom tabulations]; author's calculations

Table 6.1 Size distribution of establishments in Canada's specialized design services industry (1998–2005)

	1998				2005			
	Design services		All industries		Design services		All industries	
	#	%	#	%	#	%	#	%
1 to 4	2,450	68.2	592,981	59.2	3,271	73.2	602,086	57.2
5 to 9	546	15.2	170,797	17.1	602	13.5	178,132	16.9
10 to 19	375	10.4	115,968	11.6	359	8.0	124,486	11.8
20 to 49	193	5.4	78,690	7.9	188	4.2	90,160	8.6
50 to 99	25	0.7	24,431	2.4	36	0.8	31,819	3.0
100 to 199	3	0.1	11,154	1.1	8	0.2	15,052	1.4
200 to 499	1	0.0	5,123	0.5	3	0.1	7,576	0.7
500 or more	–	0.0	2,373	0.2	1	0.0	3,047	0.3
Total	3,593	100.0	1,001,517	100.0	4,468	100.0	1,052,358	100.0

Source: Statistics Canada, *Canadian Business Patterns 1998–2005* [author's calculations]

had fewer than five employees whereas in 2005, only 57.2 per cent of firms fell into this category.

While the growth of design activity measured in a variety of ways is impressive, there is a very explicit underlying geography to this type of activity. Design activity is primarily located in urban locations; almost 85 per cent of designers live and work in one of Canada's 27 largest cities (Table 6.2). The bulk of this employment is highly concentrated in Canada's three largest cities which act as regional centres, as well as conduits to international markets: Toronto (28 per cent), Montreal (18 per cent) and Vancouver (10 per cent). This is not surprising given the findings of a number of other studies of design and other creative activities that show that these activities are primarily located in large urban centres (Scott 2001; Florida 2002; Power 2004). By using location quotients[3], we can examine the relative importance of each location. In addition to the three largest cities, only three other Canadian cities (Victoria, Calgary, Quebec City) have a location quotient greater than one, confirming that design activity is highly concentrated in only a few urban locations. The three largest cities are also home to several prominent Canadian design schools. Furthermore, Toronto, Quebec

Table 6.2 Design employment and location quotients for major Canadian cities (2001)

Census metropolitan area	Local design labour force	% of National design labour force	Location quotient
Toronto	25,645	28.2	1.7
Montréal	16,690	18.3	1.6
Vancouver	9,120	10.0	1.5
Calgary	3,880	4.3	1.2
Québec	2,595	2.8	1.2
Victoria	1,155	1.3	1.2
Ottawa-Hull	3,555	3.9	1.0
Winnipeg	2,010	2.2	1.0
Hamilton	1,780	2.0	0.9
Halifax	1,075	1.2	0.9
Edmonton	2,480	2.7	0.8
Kitchener	1,135	1.2	0.8
London	1,075	1.2	0.8
Oshawa	775	0.9	0.8
St. Catharines-Niagara	795	0.9	0.7
Windsor	645	0.7	0.7
Saskatoon	465	0.5	0.7
Sherbrooke	325	0.4	0.7
Trois-Rivières	280	0.3	0.7
St John's	295	0.3	0.6
Kingston	250	0.3	0.6
Abbotsford	220	0.2	0.5
Chicoutimi – Jonquière	200	0.2	0.5
Saint John	170	0.2	0.5
Regina	270	0.3	0.4
Thunder Bay	130	0.1	0.4
Greater Sudbury	150	0.2	0.3
CMA	77,165	84.7	–
Non-CMA	13,935	15.3	–
Canada	91,100	100.0	1.0

Source: Statistics Canada. *Census of Population 2001* [author's calculations]

City and Victoria are provincial capitals. Calgary is a growing economic hub in the western Canada.

It is difficult to make comparisons of the size and performance of Canada's design capabilities relative to those of other national economies due to differences in data measurements and classification (Power 2004). For example, Sunley et al. (2008) report that UK design consultancies hire approximately 60,600 designers, but this excludes designers working in other industries and those who are self-employed. Power

(2004) reports that, in 2002, there were 33,768 designers in Sweden, of which only 9177 worked in the design industry. In North America, Gertler and Vinodrai (2004) found that while the United States had almost seven times the number of designers in their workforce, Canada had 4.6 designers per 1000 labour force compared to only 3.4 designers per 1000 labour force in the United States. They also found that Canadian cities had higher concentrations of designers than many of their U.S. counterparts. For example, Toronto and Montreal rank third and sixth respectively in terms of the number of designers in North American cities with a population greater than 1 million[4]. Similarly, Calgary ranked first in absolute and relative terms amongst North American cities of a similar size.

Despite the relative size of the Canadian design workforce, Canada remains unrecognized as a design nation. It should be noted that, unlike countries such as Denmark and Korea where design has been an explicit part of economic and cultural policy to bolster national competitiveness, design has not been particularly prominent within Canadian economic or cultural policy. However, while there has been no strong or coherent national strategy, the city of Montreal and the Province of Quebec have successfully incorporated design into their economic development strategies (Leslie and Rantisi 2006). It has only been very recently that regional initiatives explicitly promoting the use of (local) design have been considered in other parts of Canada (Design Industry Advisory Committee 2004; City of Toronto 2006; Mayor's Economic Competitiveness Advisory Committee 2008).

Design work in Toronto, Canada

While the locational pattern and distribution of design employment is instructive, equally interesting and useful to our understanding of design work are the organizational and territorial dynamics that underpin and reproduce this geography. I use the case of Toronto – the largest employer of designers in both absolute and relative terms in Canada – to explore these issues. According to the most recently available data, there were 2610 industrial designers and 12,680 graphic designers employed in Toronto. However, fewer than half of industrial designers (15 per cent) and graphic designers (37.5) worked in Toronto's specialized design services industry.

Global offshoring or local outsourcing?

In Toronto, it has only been recently that the policymakers have worked with design-related professional associations and institutions to understand the role of design in the local economy. This has been sparked by two related concerns. First, and following from the prevailing business and policy discourse around design and competitiveness, local policymakers and industry representatives recognized that Canadian firms face heightened global competition in international markets and that design-led strategy may help to reinvent local industries. As one design industry representative explained,

> The majority of executives know that creativity is a very powerful tool and that it might be a tool that could save them and help them become more profitable, particularly when their competition is based in Asia and [it's] on price. No North American manufacturer can win a price war. That's definitely impossible. So they're not going to win on price, so they have to win on product. To do that, they need creativity and design.

Second, there has been increased awareness amongst industry leaders that other national (and regional) governments have introduced policies geared towards fostering design activity and have strategically invested in design not just in the United States and Western Europe, but also in emerging economies (New Zealand Design Taskforce 2003, Nussbaum 2004a, City of Toronto 2006).

As flagged at the outset of this chapter, such concerns have raised the spectre amongst industry leaders and policymakers that even coveted higher wage design jobs could shift to lower cost locations outside of Canada, especially given that many of those same countries had been investing in developing design expertise (personal interviews). Similar concerns were raised by practicing designers. For example, one industrial designer described his work experience at a local firm:

> I was [at Company X] for 6 months or so and I ended up getting laid off because the manufacturer that we were working with wanted to do all of their design in-house in China and wanted to use local [Chinese] talent. So, that left me with nothing to do...a lot of design is going to end up overseas because it is cheaper to do over there.

Yet, amongst the designers interviewed for this study, there were only a few isolated instances where the designers identified similar examples of design work being relocated to offshore locations. Even the designer quoted above conceded that 'there are a lot of companies that do design in-house in Canada but then outsource all of the parts and manufacturing to China because it's cheaper.' In fact, the interviews revealed a more nuanced geography of design work that involved considerably less global offshoring of design activity compared to the prevailing views of local officials and practitioners.

Designers would 'design here' (Toronto) and negotiate the challenges around products being 'made there' (offshore locations). Design work remains highly localized in Toronto for a number of reasons. First, most design firms in Canada have a local client base and engage in the local outsourcing of design. Second, design work is project-based and precarious in nature and therefore requires a deep pool of highly skilled, local labour. Due to these conditions, designers participate in local social and knowledge networks to learn about career opportunities, as well as meet face-to-face with clients and other project members. Designers and firms – regardless of industrial context – used a suite of strategies to overcome barriers to learning and problem solving introduced by working over long distances. I discuss these dynamics in greater detail below.

'Designed here'? The localized nature of design work

Practices of firms in the design industry

There is strong evidence that design firms in Canada have sought flexibility and reduced costs by relying on freelance and contract designers (Table 6.3). The use of contract designers is highest amongst large firms (43 per cent), although larger firms account for only a small proportion of the specialized design services industry. However, even amongst small design firms, there is a high level of use of contract workers (22 per cent)[5]. Figure 6.3 further underscores this point, showing that expenditures on contract design work made by firms in the Canadian specialized design industry more than doubled between 1998 and 2002.

Table 6.3 Evidence of outsourcing relationships in Canada

	%
Design firms using contract workers[a]	
Large firms (200+)	43
Medium firms (10 to 199)	27
Small firms (<10)	22
Expenditures on contracted design work (to other firms) as % of revenue[a]	13
Self-employment amongst designers[b]	32

[a] Statistics Canada. 2002. *Annual Survey of the Specialized Design Services Industry*.
[b] Statistics Canada. 2001. *Census of Population*.

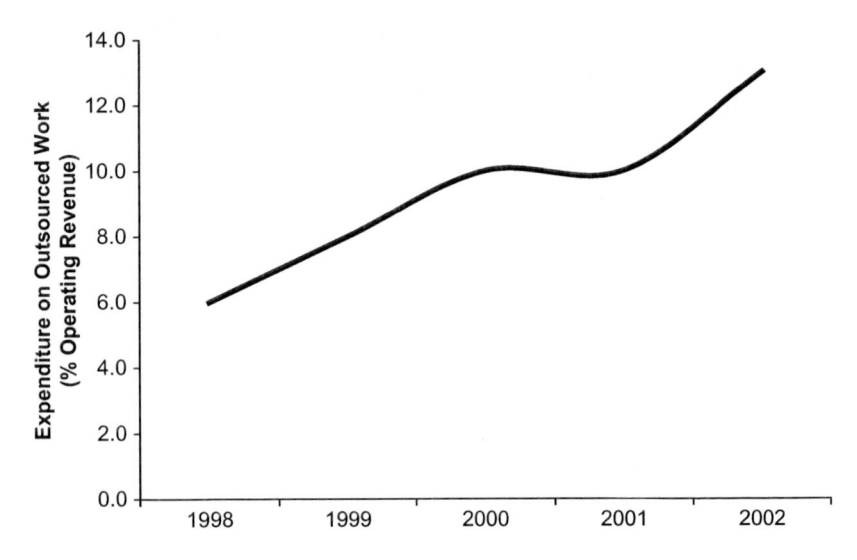

Figure 6.3 Expenditure on contract design work in Canadian firms (1998–2002)
Source: Statistics Canada. Annual Survey of Specialized Design Services, 1998–2002.

There are only limited data on firm revenue and revenue sources but evidence from Statistics Canada 2003 *Survey of Innovation* indicate that almost half of the firms in the industrial design services industry garner no revenue from international clients. Data from Statistics Canada, 2002, the *Annual Survey of the Specialized Design Services Industry*, which covers a wider spectrum of design-related industries, reveal that only 6 per cent of total revenue in the specialized design services industry is generated by international clients; the remainder is generated by Canadian businesses, the government and consumers. In other words, the design sector provides value-added services to local industries and can be viewed as an external source of innovation to local firms. Moreover, it is likely that the local design community will develop expertise that reflects the industrial structure of the particular place in which it is located.

While these data do not indicate the specific location of the designers that benefit from contract work, nor do they specify precisely where clients are located, the evidence from the interviews conducted with practicing designers and various institutional actors confirmed that most design work is contracted locally and the client base is also highly localized. The combination of these two factors provide some clues as to why design remains highly concentrated and localized even as design-related production may be (re)located elsewhere.

Project-based work, networks and local labour markets

Like many other forms of 'creative' work (Grabher 2002; Christopherson 2002), designers generally worked on project-by-project basis. As one graphic designer explained of her current work,

> It's always on a project-by-project basis. Each and every one is different. I don't always work with the same people. Sometimes I might work with [other] designers, sometimes not. I work with different business managers, client services people every time.... It is constantly evolving. I have been working on a beer product line over the last week.... I can't even remember what I did last week, things like toilet paper. Then I did a whole new coffee shop branding initiative.

While this particular designer worked in a design studio, the designers interviewed for this study reported that their work was organized on a project-by-project basis regardless of their industrial context and employment relationship (see also Vinodrai 2006). This organizational form – which provides benefits in terms of the possibilities for generating fresh and novel ideas – necessitates proximity between the project members and the client to better understand the client's needs, as well as to facilitate in solving problems that may be encountered during the design process. One designer in a manufacturing firm described his experience working with a project team and the client:

> Typically [we meet] face-to-face because we can sit down and I can sketch something out while we are talking or you know work out some ideas and then meet face to face and it is easier for me to gage somebody's reaction to things I have done when I am looking at them, when it is more interpretive, when there is more creative freedom, when there is more creative interpretation on my part in terms of what the client is looking for. It is easier to get that kind of feedback face to face.

As noted by Grabher (2002, 2004), project-based work is often short-term and project teams emerge and dissolve quite rapidly. Yet, there are a number of implications that emerge because of this dominant organizational form. The volatility of project-work and short-term contracts exposes designers themselves to high levels of precarity and risk, and they must frequently seek out new projects, work opportunities and project partners. This practice results in workers moving between projects and firms, while being tied to a particular place. As discussed elsewhere, the labour market for design work in Toronto is characterized by relatively high levels of circulation within the local labour market and most industrial and graphic designers are trained in and spend their entire career working in Toronto (Vinodrai 2006).

And while there is a strong desire on the part of designers to build a portfolio rich in experiences in different firm and industrial settings that affords designers exposure to different styles and practices, career 'disruption' is common and motivated by the precarious nature of design work itself. For example, data from the *2001 Census of Population* show that the level of self-employment amongst designers reached 32 per cent compared to only 11 per cent in the overall labour force (Gertler and Vinodrai 2004). Furthermore, more than half of the designers interviewed for this study had been laid off during their career. To mitigate the risks of this type of work, designers often engage in repeat collaborations with other local designers and project partners (Vinodrai 2006). One graphic designer at a consultancy described the decision-making process used to locate other designers and collaborators,

> 'We have lists of designers that we have worked with in the past. We have lists of designers that we know but have not yet worked with. But occasionally for a project we will have to go out and find designers that suit that project.... And they all live in [downtown Toronto].'

In other words, designers often opt to work with designers that they had previously worked with and with whom they were able to work effectively and trust. Such relationships provide the promise of future work and belonging to well-developed social networks provides a better guarantee at future employment. Even when designers would seek out new project and work partners through local search, this search would also be is facilitated through participation in local social networks. One designer explained her rationale for relying on her own local social networks,

> I have a social network here.... The designers we work with in Toronto are more likely to hit a home run because they're more likely to talk to us or they call and ask us questions or they feel they know us better.... So in some ways Toronto has a higher batting average

[compared to] the people we work with in other cities... It's not that they are more or less talented than Toronto designers – but... it takes more effort to maintain a relationship over distance.

While these social dynamics are not surprising, they confirm the findings of other studies of project-based creative work in a variety of other geographic, industrial and occupational settings and show that designers become embedded and grounded in particular social and networks tied to particular places (Christopherson 2002, Grabher 2002, 2004; Reimer et al. 2008).

'Made there'? Overcoming the proximity challenge

While there is strong evidence that design work itself has remained strongly rooted in Toronto, there are some challenges associated with 'being there', yet not being proximate to production facilities (Gertler 1995, 2004; Gertler and Vinodrai 2005). Two common challenges were cited by practicing designers. First, a number of designers expressed frustration with the interpretation of their designs once they were in the hands of offshore manufacturers. One designer tried to explain, 'They are just coming from a different place. So, what's intuitive to you about the product is not intuitive to them. So like you've just got to imagine that someone knows absolutely nothing about [Product X]. And you had to label every little bit. You know, make sure that it can't be like that, it has to be like this.' Second, a number of industrial designers expressed concerns that the cost-reducing motives for seeking offshore partners in the first place hampered their ability to do better design work and that there were often significant technological limits and capabilities at the offshore manufacturing sites that the designers were often not aware of at the outset of a project. As one designer put it,

> A lot of the limits that we have here are related to [offshore] production. It's not cutting edge. The molders that we use are not the leaders. For example, the level of detail for an injection mold part in terms of the tolerances and the quality and the fine details and stuff – you can't achieve a lot of that because you need to use an expensive mold.

Yet, firms inside and outside of the design industry were able to negotiate this lack of cultural and technological understanding and proximity through a number of different strategies. Designers explained that once they were able to learn about the limits of their offshore partners, they were better able to design products that fit within the technological capabilities of the offshore facility. A number of designers identified strategies for learning and understanding their distant project partners' capabilities. As one designer explained,

> Most of my designer friends who work with China, they go to China like once a year, twice a year to go see and to go look or whatever and the big companies that have major operations in China had a China office and they might employ an industrial designer full-time there.

At the firm level, a number of firms were able to hire designers and managers with regional experience/expertise or a cultural affinity to the regions in which their offshore facilities were located to overcome some of the language barriers involved in interpreting and translating the designs themselves. As one designer working for an equipment manufacturer explained,

> [The designs] go through [my manager] partially because he speaks Chinese and I don't. Partially because he tries to keep a very tight reign on the information that goes to or from this building and China. The main thing is he's trying to keep cost in check and mistakes from happening so essentially everything that goes in and out of this building in terms of manufacturing in China goes through [my manager]

Other firms engaged the services of a third-party mediator to both seek out offshore locations and then to work with the Toronto-based designers and project team, as well as the offshore manufacturer and – increasingly – their local designer (personal interviews).

Working here, made there? Conclusions and policy implications

Overall, this chapter has shown the growth in design activity in Canada and has further confirmed that this activity is highly localized in major urban centres. This suggests that when considering the relationship between national competitiveness and design, it is imperative to understand design work as taking place in large, urban centres located within broader global production networks. The case of Toronto, Canada is instructive in this regard. Certainly, in Toronto, the highly localized nature of design work presents both opportunities and challenges to designers, the users of design services and policymakers. The localization of design work in Toronto highlights the spatial disconnect between creative work taking place in cities in open, advanced economies and the downstream activities related to production at (often) geographically distant sites. This perceived spatial division of labour has raised some anxiety and fear amongst Canadian practitioners and policymakers alike who are concerned about the potential outsourcing and offshoring of design activity to be closer to these 'offshore' sites of production.

However, the evidence reviewed herein suggests that these fears are somewhat misplaced and overstated. While true that there has been some outsourcing of design work in Toronto, more often than not this design work has been contracted out to *local* freelancers rather than to offshore locations. Furthermore, to the extent that firms in Canada are shifting production to lower cost locations, there are strong arguments in favour of maintaining the (creative) design work in Canadian cities where they are close to their clients and the market they serve. Rather than relocating design work alongside production work, firms seek out alternative ways to overcome the proximity challenges presented by operating in multiple geographic contexts, including hiring individuals with experience and knowledge of the production location (shared social, linguistic and cultural identity) and involving designers in both locations (shared occupational identity). In other words, we may be witnessing a simultaneous globalization and localization of design activity.

The continued localization of design work within large urban centres presents a number of opportunities for policymakers, particularly in open, advanced national economies. The findings of this study suggest that promoting and supporting the local design community and maintaining design jobs in major urban centres can be a means of bolstering both regional and national competitiveness. First, design work remains embedded within the cities of advanced, open national economies due to the very nature of design work. Design work is project-based and requires the mobility of designers between different projects (or firms), and this is best achieved through thick local labour markets that strongly root (creative) design work to particular cities and nation-states. Furthermore, the presence of a large pool of designers affords firms inside and outside of the design industry ready access to local design expertise. Second, due to designers' relatively high levels of labour market mobility between firms and the project-based nature of design work, design skills can be transferred across industry sectors, thereby opening the door for cross-sector knowledge transfer and innovation. Third, and related, this mobility between firms and projects provides an opportunity to strengthen existing local networks, as well as bridge global networks through interactions with both local and non-local clients and designers. In fact, as production networks continue to evolve, it is more likely that more highly-developed and robust global design networks will emerge, resulting in the transfer of local design knowledge between cities in open, advanced nations and offshore locations. Such global 'pipelines' are suggested to be vital sources of new knowledge and ideas that help to maintain the innovative dynamism and competitiveness of local and national economies (Bathelt et al. 2004).

Overall, this suggests policymakers should encourage the use of design, especially in 'sticky' industries that are likely to be the source of on-going employment in the local or national economy, as well as promote the use of design in 'traditional industries'. In this way, local design expertise can be harnessed to facilitate the process of industrial upgrading and improve the position of firms, regions and nations in the global marketplace. In other words, the very practice of 'designed here' provides a possible means to anchor some industries, potentially slowing the 'made there' phenomenon and improving the economic competitiveness of firms, regions and nations. However, this remains an open theoretical and empirical question best addressed by further research.

Acknowledgments

An earlier version of this paper was presented at the special session 'Offshore, Onshore, Nearshore, and Blended-Shore: Creative and Cultural Industries' at the Annual Meeting of the American Association of Geographers, Chicago, Illinois, 8March 2006. The author would like to thank Meric Gertler and Nichola Lowe for their helpful comments and suggestions, Deborah Leslie and the Cultural Economy Laboratory at the University of Toronto, the Design Industry Advisory Committee and the City of Toronto Economic Development Office for access to resources and data and the Social Science and Humanities Research Council for funding the research upon which this chapter is based. The author would also like to extend thanks to the editors for all of their hard work in putting together this collection.

> It is no longer just manufacturing that is feeling the pressure of foreign competition. It is no longer just dirty blue-collar jobs that are moving offshore. Jobs in services are now migrating as well, some of them requiring advanced skills…. Services constitute much the

larger part of every advanced economy. At the end of this process, what will be left? (*The Economist*, 21–27 February 2004, 11)

What comes after services? Creativity. (*Wired*, February 2004, 99)

In a world enriched by abundance but disrupted by the automation and outsourcing of white-collar work, everyone regardless of profession must cultivate an artistic sensibility. We may not all be Dali or Degas. But today we must all be designers. (Daniel Pink, A Whole New Mind: Moving from the Information Age to the Conceptual Age)

Notes

1. Design is defined in occupational terms to include industrial designers, graphic designers, interior designers, theatre, exhibit and other designers, as well as architects and landscape architects. Design is defined in sectoral terms using the North American Industrial Classification System (NAICS) as the specialized design services industry (5414) and can include workers in a variety of occupations, including design. Missing from both of these measures are individuals who carry out design work but hold other job titles and/or persons who only conduct design-related activities as one aspect of their work.
2. The specialized design services industry had a compound annual growth rate of 2.8 per cent compared to 0.6 per cent for the Canadian economy as a whole.
3. Location quotients are a measure of relative size (or specialization). In this case, the location quotient compares the proportion of a city-region's employment accounted for by designers to the national average. A value greater than 1 suggests that the city-region has a higher-than-expected level of design employment and indicates that the city-region has specialization in this activity.
4. Toronto also ranks third in relative terms (adjusted for population) and Montreal ranks fourth. Due to data constraints, all comparisons exclude self-employed designers.
5. Unfortunately, data on the actual locations of the contracted workers is not available.

References

Amin, A. and Cohendet, P. 2004. *Architectures of knowledge: Firms, capabilities, and communities.* Oxford University Press, Oxford.
Bathelt, H., Malmberg, A. and Maskell, A. 2004. 'Clusters and knowledge: Local buzz, global pipelines and the process of knowledge creation', *Progress in Human Geography* 28, 31–56.
Bryson, J.R., Daniels, P.W., Warf, B. 2004. Service Worlds: People, Organisations, Technologies, Routledge, London

Christopherson, S. 2002. 'Project work in context: Regulatory change and the new geography of media', *Environment and Planning A* 34, 2003–15.
City of Toronto. 2006. *Drawing the link: Advancing design as a vehicle for innovation and economic development*. City of Toronto Economic Development Division, Toronto.
Coe, N.M. and Bunnell, T.G. 2003. '"Spatializing" knowledge communities: Towards a conceptualization of transnational innovation networks', *Global Networks* 3, 437–56.
Core 77 Design Network. 2002. *Industry snapshot: A primary research study covering demographics, business characteristics, growth strategies and best practices within the design industry*. Core 77 Network, New York.
Currid, E. 2007. *The Warhol economy: How fashion, art and music drive New York City*. Princeton University Press, Princeton, NJ.
Danish Design Centre. 2003. *The economic effects of design*. National Agency for Enterprise and Housing, Copenhagen.
Design Council. 2004. *Design in Britain, 2004–2005*. (http://www.design-council.org.uk).
Design Exchange. 1995. *Design for a strong Ontario: A strategy for Ontario's design sector*. Design Exchange, Toronto.
Design Industry Advisory Committee. 2004. *What can 40,000 designers do for Ontario? Design matters*. Design Industry Advisory Committee, Toronto.
Ekinsmyth, C. 2002. 'Project organization, embeddedness, and risk in magazine publishing', *Regional Studies* 36, 229–44.
Economist, The. 2004. 'The new jobs migration: Foreign competition now affects services as well as manufacturing'. *The Economist* 370, 11.
Feser, E.J. 2003. 'What regions do rather than make: A proposed set of knowledge-based occupation clusters', *Urban Studies* 40, 1937–58.
Florida, R. 2002. *The rise of the creative class*. Basic Books, New York, NY.
Gertler, M.S. 1995. 'Being there: Proximity, organization, and culture in the development and adoption of advanced manufacturing technologies', *Economic Geography* 75, 1–26.
Gertler, M.S. 2003. 'Tacit knowledge and the economic geography of context, or the undefinable tacitness of being (there)', *Journal of Economic Geography* 3, 75–99.
Gertler, M.S. 2004. *Manufacturing culture: The institutional geography of industrial practice*. Oxford University Press, Oxford.
Gertler, M.S. 2008. 'Buzz without being there? Communities of practice in context'. In Amin, A. and Roberts, J. (eds) *Community, economic creativity, and organization*. 203–26. Oxford University Press, Oxford.
Gertler, M.S. and Vinodrai, T. 2004. *Designing the economy: A profile of Ontario's design workforce*. A report to the Design Industry Advisory Committee, Toronto.
Gertler, M.S. and Vinodrai, T. 2005. 'Learning from America? Knowledge flows and industrial practices of German firms in North America', *Economic Geography* 81, 31–52.
Grabher, G. 2002. 'The project ecology of advertising: Talents, tasks, and teams', *Regional Studies* 36, 245–62.
Grabher, G. 2004. 'Learning in projects, remembering in networks? Communality, sociality, and connectivity in project ecologies', *European Urban and Regional Studies* 11, 103–23.

Kotler, P. and Rath, G.A. 1984. 'Design: A powerful but neglected strategic tool', *Journal of Business Strategy* 5:2, 16.
Lash, S. and Urry, J. 1994. *Economies of signs and spaces*. Sage, London.
Leslie, D. and Rantisi, N. 2006. 'Governing the design economy of Montreal, Canada', *Urban Affairs Review* 41, 309–37.
Leslie, D. and Reimer, S. 2006. 'Situating design in the Canadian furniture industry', *The Canadian Geographer* 50, 319–41.
Lloyd, R. 2006. *Neo-bohemia: Art and commerce in the post-industrial city*. Routledge, New York, NY.
Lorenzen, M. (ed.) 1998. *Specialization and localized learning: Six studies on the European furniture industry*. CBS Press, Copenhagen.
Markusen, A. 2004. 'Targeting occupations in regional and community economic development', *Journal of the American Planning Association* 70, 253–68.
Martin, R. 2006. 'What innovation advantage?' *Business Week* 16 January.
Maskell, P. and Malmberg, A. 1999. 'Localised learning and industrial competitiveness' *Cambridge Journal of Economics*, 23, 167–85.
Mayor's Economic Competitiveness Advisory Committee. 2008. *Agenda for prosperity*. City of Toronto, Toronto.
Merritt, J. and Lavelle, L. 2005. 'Tomorrow's B-School? It might be A D-School'. *Business Week* 1 August.
Molotch, H. 2002. 'Place in product', *International Journal of Urban and Regional Research* 26, 665–88.
New Zealand Design Taskforce. 2003. *Success by design: A report and strategic plan*. A report prepared in partnership with the New Zealand government.
New Zealand Institute of Economic Research. 2003. *Making the case for added value through design*. Report to Industry New Zealand.
Nussbaum, B. 2004a. 'Redesigning American business'. *Business Week* 29 November. (http://www.businessweek.com/bwdaily/dnflash/nov2004/nf20041129_2629.htm).
Nussbaum, B. 2004b. 'The power of design'. *Business Week* 17 May, 86–94.
Owen-Smith, J. and Powell, W.W. 2004. 'Knowledge networks as channels and conduits: The effects of spillovers in the Boston biotechnology community', *Organization Science* 15, 5–21.
Pink, D. 2005. *A whole new mind: Moving from the information age to the conceptual age*. Riverhead Publishers, New York, NY.
Power, D. 2002. '"Cultural industries" in Sweden: An assessment of their place in the Swedish economy', *Economic Geography* 78, 103–27.
Power, D. 2004. *The future in design: The competitiveness and industrial dynamics of the Nordic design industry*. Centre for Research on Innovation and Industrial Dynamics, Uppsala.
Power, D. and Scott, A.J. (eds) 2004. *The cultural industries and the production of culture*. Routledge, London.
Pratt, A.C. 1997. 'Employment in the cultural industries sector: A case study of Britain, 1984–91', *Environment and Planning A* 29:11, 1953–76.
Rantisi, N.M. 2002. 'The competitive foundations of localized learning and innovation: The case of the women's garment industry in New York City', *Economic Geography* 78, 441–61.

Rantisi, N.M. 2004. 'The designer in the city and the city in the designer'. In Power, D. and Scott, A.J. (eds) *The cultural industries and the production of culture*, 91–109. Routledge, London.

Rantisi, N.M. and Leslie, D. 2006. 'Branding the design metropole: The case of Montréal, Canada', *Area* 38:4, 364–74.

Reimer, S., Pinch, S. and Sunley, P. 2008. 'Design spaces: Agglomeration and creativity in British design agencies', *Geografiska Annaler, Series B: Human Geography* 90, 151–72.

Rusten, G. and Bryson, J. 2007. 'The production and consumption of industrial design expertise by small and medium-sized firms: Some evidence from Norway', *Geografiska Annaler, Series B: Human Geography* 89, Issue S1, 75–87.

Scott, A.J. 2001. *The cultural economy of cities: Essays on the geography of image-producing industries*. Sage Publications, Oxford.

Scott, A.J. 2004. 'Cultural-products industries and urban economic development: Prospects for growth and market contestation in global context', *Urban Affairs Review* 39, 461–90.

Statistics Canada 2002. *Annual Survey of the Specialist Design Services Industry*, Statistics Canada, Ottawa, Ontario.

Statistics Canada 2003. *Survey of Innovation*, Statistics Canada, Ottawa, Ontario.

Storper, M. and Venables, A.J. 2004. 'Buzz: Face-to-face contact and the urban economy', *Journal of Economic Geography* 4, 351–70.

Sunley, P., Pinch, S., Reimer, S. and Macmillen, J. 2008. 'Innovation in a creative production system: The case of design', *Journal of Economic Geography* 8, 675–98.

Vinodrai, T. 2005. *Locating design work: Innovation, institutions and local labour market dynamics*. Unpublished doctoral dissertation. Department of Geography, University of Toronto.

Vinodrai, T. 2006. 'Reproducing Toronto's design ecology: Career paths, intermediaries, and local labor markets', *Economic Geography* 82:3, 237–63.

Wired. 2004. 'Why coders and call centers are just the beginning'. *Wired* February, 99.

7
The Geography of Design in the City
Thomas A. Hutton

Introduction: Design, the cultural economy and the 'resurgent city'

The rise of a cultural economy and its constituent industries, institutions and labour has generated a new phase of growth in the metropolis, underpinning what Allen Scott has termed the 'resurgent city' in the early 21st century (Scott 2007). While this experience of culture-led growth is jurisdictionally selective, and generates an uneven distribution of benefits and costs, it does appear to offer a robust platform for economic development across a growing number of cities among advanced and transitional societies, although its universal potential is routinely exaggerated. There are also definitional issues to consider, as the cultural economy comprises complex systems of production and consumption, and features strong interdependencies with residential markets and urban lifestyles (see Zukin 1989, 1998 in this connection). That said, central to each of these facets of the cultural economy is the concept of design: as a defining attribute of specialized production; as a marker of consumption and social class distinction; as a shaper of contemporary urbanism and as a key characteristic of housing preference among affluent social actors, notably the 'creatives' and others within the new middle class (Hamnett and Whitelegg 2007).

Investigations of the growth dynamics of the cultural economy stress a range of causal factors and interdependencies. These include (on the supply side) the reassertion of specialized production within the terrains of the post-industrial city, shaped by a recovery of the city's creative vocation and its constitutive concentrations of skilled labour, infrastructure and institutions (see Hutton 2008: 2039). Demand factors include both intermediate demand (for example, among industries engaged in the fabrication of 'cultural products' – services as well as goods) and final demand (as in the evolving preferences of affluent consumer markets for cultural goods which signify identity; see Scott 1997; Evans 2003). Together these factors underpin the emergence of the cultural economy as an important phase of post-Fordism, embodying a complex range of social consequences as well as the ascendancy of a distinctive industrial production régime (Power and Scott 2004).

The cultural economy encompasses long-established industries such as architecture, industrial design, advertising and corporate branding, as well as those associated with the New Economy innovations of the 1980s and 1990s, a trajectory which includes, for example, software design, digital arts, computer graphics and imaging, video game production, and other forms of new media. These industries are complemented by ancillary enterprises (such as technical support firms) and a system of intricate input-output relations operating within localized production networks, forming 'new industrial districts' in the heart of the city and in some of the peripheral areas of the metropolitan region. Increasingly the development of the cultural economy of the city is fostered by a host of supportive institutions and programs, including higher education, training and apprenticeships, as well as heritage, land use and marketing policies. The cultural economy also comprises a profusion of consumption activities which contribute to the conviviality and spectacle of the contemporary city. These are in turn associated with a new urban aesthetic and lifestyle, and with insistent place-making (and marketing), generating in turn a significant cachet for contemporary culture in the media and in public discourses. But a concept quite central to each of these diverse cultural activities is that of design, a term with a more precise connotation than that of 'creativity', the utility of which has been somewhat compromised by promiscuous usage and expression.

'Design' carries with it a manifestly industrial application, both in its contemporary and more historical applications (Hutton 2000). Design processes include not only the specification of the physical attributes of product development (including product morphology and articulation of constituent elements) across the full range of consumer and capital goods, but also the critical conceptual linking of demand, innovation and production, and the selection of inputs for fabrication: material resources, labour and technology. Its lineage includes the increasing centrality of design to the advanced industrial economies of 19th century Europe, initially in England and then in other northern European states, and indeed encompasses a more extended reach back to the artisanal industries of the pre-industrial city. As industrialization progresses to more advanced stages, the role of design becomes increasingly crucial, not only in the pursuit of 'style' and other aesthetic embellishments, but also in the exigent tasks of achieving greater production efficiencies and economies of resource utilization, contributing to the professionalization of design as a discipline and enterprise.

The centrality of design to advanced industrial systems also has a powerful occupational correlate, in the evolution of key task specializations and social and technical divisions of production: from artisanal labour in the period of craft production; to proto-industrial labour in the early years of the industrial revolution; then to the burgeoning industrial labour, working class and community formation of the 19th and early 20th centuries; followed by the innovations of Fordist production technology and Taylorian labour which typified the advanced industrial economies of the mid-20th century; and then, and more recently, to the emergence of post-Fordist (or neo-artisanal) labour engaged in design work among the cultural industries (Norcliffe and Eberts 1999).

As for earlier phases of economic development, including successive industrialization stages and the rise of the service economy, there is a distinctive multiscalar geography which characterizes the cultural economy, and more specifically the design functions which constitute its motive force. First, we can readily discern a spatiality of design expressed at the regional scale, exemplified in the cultural product fabrication firms situated within the 'new industrial districts' of the 'Third Italy' (Becattini 1990), including the intricate, design-intensive production networks of Emilia-Romagna and the Veneto, and the fashion industry of the Île de France (Storper and Salais 1997). Secondly, there are the multiple associations between specific cities and particular high-value, high-design product sectors, such as Florence (leather goods), Milan (fashions), Venice (glass), Paris (textiles), Geneva (watch-making), Leipzig (book-publishing), Helsinki (furniture design) and so on (Hall 1998). Finally, at the more intensely localized scale, the emergence of design industry clusters has been acknowledged as an integral element of the contemporary cultural economy, exhibiting in some respects historical continuities (for example, precision trades in Clerkenwell, London; see Evans 2004) but in others significant departures from past patterns and practices, as in the case of Leipzig (Bathelt and Boggs 2003). But the geography of design is becoming more complex, and increasingly cannot be tidily arrayed within these discrete spatial units, as insistent outsourcing, enabled by the Internet and other forms of telecommunications, reconfigures the scope of cultural production networks and systems.

My purpose in this chapter is to offer a profile of the geography of design as a key feature of the contemporary cultural economy of the city, stressing the saliency of place as well as space, and acknowledging the mix of development factors and interdependencies influencing location. Following this introduction, the chapter situates design in the city, and then advances a model of design activity at the urban-regional scale. This model acknowledges logics of industrial location, as well as more recent efforts to depict the spatiality of cultural production in the metropolis which give greater weight to social, environmental and policy factors. Next, a set of illustrative examples of key sectors and instructive case studies at the city and district level will demonstrate the rich empirical dimensions of design activity, informed by this earlier conceptual enterprise and by an extensive program of field work, and including, primarily, sketches of design industry formation in two instructive cities: London and Vancouver. The discussion will include a description of the tensions produced by cost pressures and burgeoning outsourcing tendencies, the development of more extended production networks and global recruitment of key labour cohorts, which modify the operating characteristics of design industries within localized clusters. A concluding section will summarize key findings and suggest some new directions for research in this lively domain of inquiry which synthesizes economic geography, urban studies and regional development.

Interdependencies of design, innovation and place in the city

Design was crucial both to the pre-industrial (artisanal) activity of the city in history as well as to the advancement of industrial production modes among the leading 19th century economies. But the significance of design to the contemporary cultural economy and to the progression of advanced industrial economies more generally has been to an extent occluded, both by the uncritical treatment of 'creativity' as a marker of leading-edge economic change, and also by the sequence of abbreviated restructuring episodes which have to some extent subverted the task of effective retheorization.

In this regard Allen Scott (2007) has identified no fewer than seven descriptors of changes since the collapse of Fordism among advanced economies since the 1980s, including the so-called 'sunrise' and 'sunset' industries, the emergence of the 'knowledge-based economy' (KBE) and the technology-driven 'New Economy' of the 1990s and the contemporary cultural economy with its associated creative industries, labour and social class manifestations. Scott acknowledges the explanatory limits of post-Fordism as model of economic development amid these volatile episodes of innovation and restructuring, and proposes as a successor concept the 'cognitive cultural economy' to encompass the complexities of industrial organization and labour formation (Scott 2007). In this interpretation a synthesis of cultural production activities and labour reflects a new phase of flexible specialization among advanced economies, and offers possibilities for growth for cities in the aftermath of secular decline in Fordist manufacturing and (in some cases) mainstream service industries and employment.

Design constitutes a defining attribute of economic development and also the transformation of cities and society. Design in the broadest sense is deployed increasingly in the generation of industrial products and services, in the production of food and other high-value consumption items and in the construction of social identity and lifestyles. Cities themselves have become objects as well as sites of design, as elucidated in Harvey Molotch's well-known evocation of Los Angeles as a design product (1999), and in the 'creative city' movement promoted by Richard Florida and others, in which all aspects of life – work, recreation and leisure – can ostensibly be enhanced through the value-added potential of culture and design (Florida 2002). Reflecting this exaltation of the post-Fordist 'designer city', urban design occupies an increasingly privileged place among the repertoire of local policy lexicons and instruments, relative to more prosaic spheres such as land use regulation and development control, constituting a vital component of contemporary place-making (and marketing). And while housing has always been a principal marker of social status and class identity, affluent consumers such as the supergentrifiers (Butler and Lees 2006) and members of the transnational élites are increasingly turning to architects and other design professionals (such as interior designers, landscape architects) to make a cultural statement via design, contributing to the reconstruction of place, as well as to identify formation.

A spatial model of design activity in the metropolis

While for each of these aspects of 'cultural urbanism' design functions are clearly integral to their formative processes, I am principally concerned here with the role of design in the production relations of contemporary industrialization, and more particularly with the place of design in the evolving space-economy of the metropolis. In this realm, as for other centrepieces of the urban-regional economy, we can readily discern the spatial lineaments of design activity, shaped by economic agglomeration forces, as well as by social and environmental factors associated with the cultural economy and its constituent industries and labour.

Figure 7.1 depicts a schematic representation of design industries positioned with the zonal structure of the metropolis. In general terms this model follows Allen Scott's model of industrial location in the urban region, first, in the tendency for smaller, more contact-intensive operations to congregate within the central and inner city, and for larger, more self-contained enterprises to situate in more peripheral areas (Scott 1982), and, secondly, in the ever-more fine-grained industrial specialization of the internal spaces of the metropolis (Scott 1988). The pervasiveness of specialized design activity throughout the differentiated

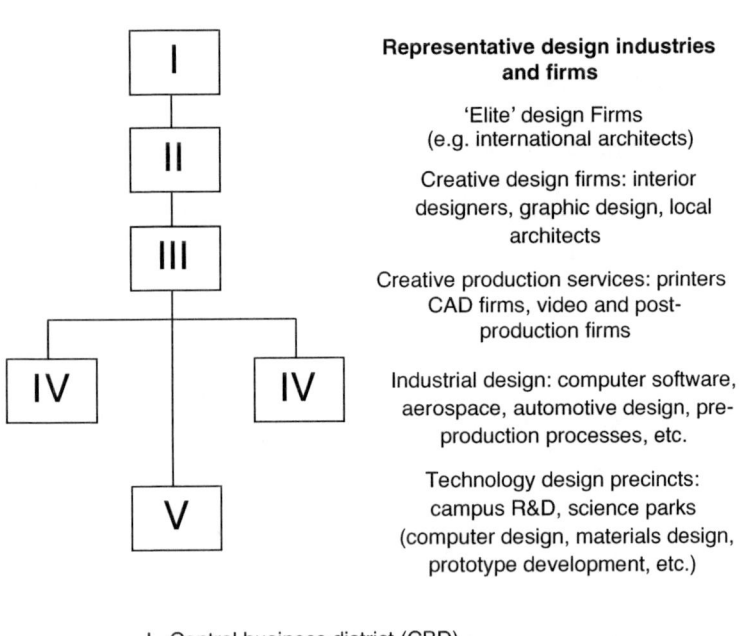

Representative design industries and firms

'Elite' design Firms
(e.g. international architects)

Creative design firms: interior designers, graphic design, local architects

Creative production services: printers CAD firms, video and post-production firms

Industrial design: computer software, aerospace, automotive design, pre-production processes, etc.

Technology design precincts: campus R&D, science parks (computer design, materials design, prototype development, etc.)

I   Central business district (CBD)
II CBD fringe and inner city
III Inner city and mid-town precincts
IV Inner/outer suburbs
V   Metropolitan fringe and 'edge city'/exurban sites

Figure 7.1 Schematic of applied design clusters within the metropolitan space economy

economic landscapes of the metropolitan space-economy is suggestive of its importance to the development trajectory of the city, and to the centrality of design to the core industrial functions and labour specializations among the most advanced urban societies.

To elaborate upon the principal locational characteristics of design firms, Figure 7.1 incorporates the major clusters of design activity concentrated within industrial clusters of the metropolis, including both the Central Business District (CBD) and more particularly in the post-industrial landscapes of the CBD fringe and inner city. The modernist office towers of the CBD accommodate 'élite' design firms in important cultural industries, including international architects and landscape architects, prominent graphic designers, and industrial designers. To an extent these CBD-based design operations are engaged in corporate control and management, particularly in firms with extensive international practices, but there is also significant production activity conducted in studios and office space within the central city. Furthermore, over the last decade or so the CBD has accommodated some of the edgier design industries, including video game production, exemplified in an operation of Electronic Arts situated in Vancouver's central office complex. Here, as more and more routinized or back office functions are decentralized to more peripheral sites within the region or (increasingly) beyond, the vacated space can be 'recolonized' by high-value industries in the cultural economy sector, representing a new phase of succession in the city.

But in many cities the largest concentrations of design industries and firms are situated within the heritage landscapes and buildings of inner city districts, with prominent examples including London's 'City Fringe' (incorporating Clerkenwell, Shoreditch, Soho and Bermondsey Street, among other districts), the Biccoca district of Milan (Sacco and Tavano Blessi 2008), Lower Manhattan in New York (Indergaard 2004), the South of Market Area (SOMA) in San Francisco (Hutton 2008: 178–221), and South Melbourne (Elliott 2005). In these districts we typically find major constellations of design-based industries and firms, including, foremost, industries dependent upon a fluid synthesis of culture, social capital, technology and attributes of place in specialized production. In the largest cities, the representation of design-based industries and firms encompasses extraordinary diversity across the principal production régimes, including dynamic post-Fordist industries such as new media, software development and film and video production; intermediate service industries imbued with significant design content, notably advertising; 'retooled' Fordist industries such as printing and publishing and a range of pre-Fordist 'relics', such as instrument makers, custom woodworkers, bookbinders and others.

For some critical studies scholars, notably Marxists, this affiliation between creative industries and the spaces of the inner city can be reduced to the actions of the property market. This viewpoint is characterized by a storyline of predictable declines in property values ensuing from the collapse of traditional industry in the core, followed by an insistent revalorization shaped initially by artists and by gentrifiers, and then by subsequent episodes of succession and dislocation driven by ascendant social cohorts in the city. These cohorts include (on the demand side) gentrifiers as well as property market actors, such as investors, developers and estate agencies.

While the behaviour of the urban property is indeed implicated in successive rounds of land use change in the core, though, the concentration of design-based industries within the landscapes of the metropolitan core is more directly associated with the larger reassertion of specialized production in the inner city. This reinsertion of industry in the core is underpinned by a broader range of factors which include: (1) proximate skilled labour supply, and more specifically, the co-presence of artists and designers which are incorporated within the intricate articulations of cultural production within the urban core; (2) the unique physical and representational values of the inner city built environment; (3) the distinctive spatiality of the urban core; (4) the unique 'institutional thickness' of the central city, including galleries and exhibition space as well as schools of art and design and (5) the rich amenity base of the central and inner city, which serves to attract creative industries and labour, and which lubricates the social interaction which typifies the behaviour of many design-oriented industries and firms. While (as we shall see) there are cost pressures and increasing centrifugal forces acting on the cultural industries of the core, these conditions in the aggregate constitute a powerful magnet for many design industries, institutions, firms and labour. The urban core, including the CBD and inner city, constitutes both the traditional heart of the metropolis and the essential locus of establishment and operation for the design-based industries which primarily shape the development of the cultural economy of the metropolis.

That said, there are significant – and in many cases rapidly-growing – representations of design-based industries, institutions and activities situated within more peripheral zones of the metropolis, demonstrating logics of location which increasingly feature technology both directly as inputs to production and as channels connecting firms in extended production networks, as well as underscoring the increasing penetration of cultural production within the urban-regional space-economy. The growth of advanced design firms and other cultural industries represents a further development and maturation of suburban (and in cases exurban) economies.

To an extent the location of these design-based firms beyond the confines of the urban core reflects a requirement for larger sites and/or buildings than those easily obtained in the urban core. The need for larger premises among certain design firms may be associated with the scale of enterprise and production, the need for integrated on-site design and fabrication and/or the scalar dimensions of plant and equipment. Examples include film production studios, which may be situated within inner suburban sites which offer both larger properties as well as relatively favourable access to the central city; industrial research and design operations embedded within suburban universities and science parks; and aerospace industries which may require space-extensive sites in outer suburban or exurban areas for prototype design and testing. Design-intensive industries can in some cases locate within industrial parks, taking advantage of local service provision and (generally) cheaper rents than those which prevail within the metropolitan core. Examples here include the Mexx design centre located in the outer zone of Amsterdam, on the road to Schiphol; and film production activity situated within the inner suburban districts (notably Burnaby and North Vancouver) of Vancouver. These larger concerns can in many cases operate somewhat independently of the specialized services and ancillary industries of the urban core.

Cities, sites and the geography of design

The location of design firms in the metropolis reflects the spatial organization of specialized production among advanced economies, and embodies both a layering of arts and design traditions, as well as features of the sequences of industrial innovation and restructuring. This accretion of arts and design capacity within the spaces of the city can include, variously, centuries-old traditions of fine arts and craft production (Florence, Hanoi), specialized industries and skill sets associated with the artisanal production of 19th century industries economies (Clerkenwell, Manhattan), the expansion of artists' communities in post-industrial districts of post-Fordist economies (Paris, London, Prague), and the more recent burgeoning of contemporary design activity as an evocation of the post-Fordist cultural economy of the city.

These primary design activities all form part of the critical capital stock of cultural production, but the inherent tensions and instabilities which characterize their relations, and their uneven articulation within industrial production systems, may also be acknowledged. Andy Pratt has described a process of 'industrial gentrification' which acts to squeeze lower-margin cultural firms and labour in the city, in the face of competition from more profitable creative firms and mainstream financial-commercial business (Pratt forthcoming), while at the same time professionalization within the cultural sector tends to privilege credentialized workers across a range of occupations (Ho 2007). The inflationary housing markets of the city, shaped in part by the aestheticisation of place, increasingly displace younger cultural workers and start-up firms. And for even the most successful of the cultural industry workforce the exigent pressures of competition, deadlines and long and often irregular hours of work place enormous stress on family life and other social relations.

To an extent the industrial manifestations of design activity can be seen as expressions of structural forces, evidenced by a proliferation of design-based industries both in Western and 'transitional' economies, including those of East and South-east Asia; but local contingencies (for example, urban scale, economic histories and legacies, culture and governance) have the power to reshape global processes at the micro-scale, producing significant differences in the specific composition of industries, firms and labour. What follows, then, is an elucidation of design activities situated in two instructive cities, London and Vancouver, narratives which include an acknowledgement of both structural and more locally-specified features.

London: design in the central area and 'city fringe'

The scale and specialization of design activity in London reflects in full the capital's global city status and vocations, with arts and applied design industries contributing to the cluster of banking, finance, business services and political power which form the foundations of its international market presence and projection, encompassed within Peter Hall's model of contemporary global city service functions (Figure 7.2). As for other key sectors, London is both a site and agency of specialized design production, including world-scale concentrations of firms practicing in industrial, fashion, and graphic design; influential architects (Norman Foster, Richard Rodgers) and landscape architects; advertising and corporate branding; printing and publishing and new media. Much of this design activity has been deployed in the reshaping of London

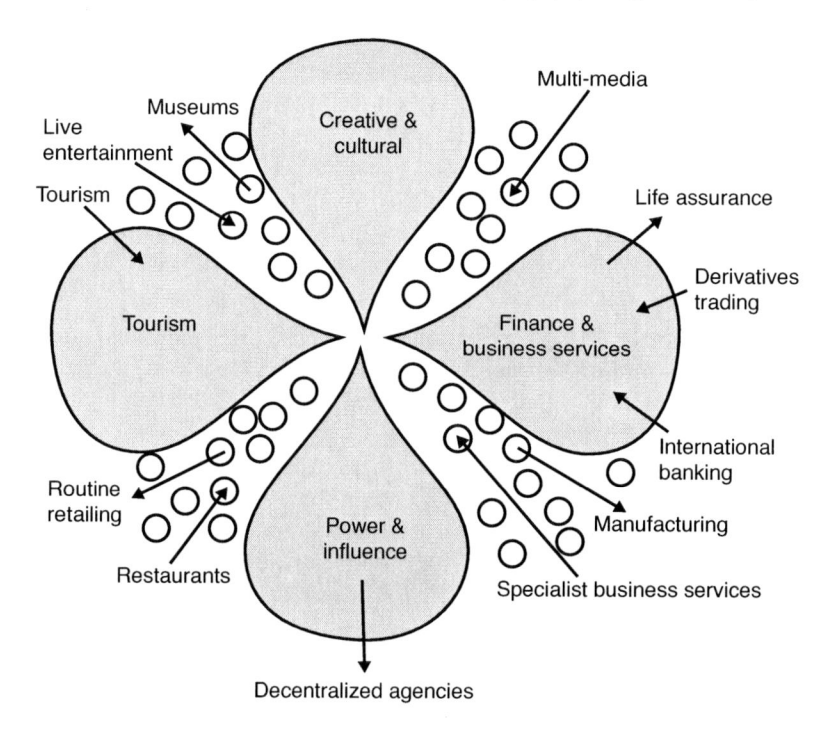

Figure 7.2 Major service clusters in the polycentric global city
Source: Davies and Hall 2006

itself, an integral feature of its resurgence since the early 1990s, as seen in its new architecture and built form, iconic buildings, new public spaces and sites of spectacle and consumption, public art and derived imageries and place-making (Judt 2005; Kaika and Thielen 2005).

There are concentrations of elite designers domiciled within the City of London, the most prestigious business location in the metropolis, including architects, corporate branding and advertising and at least a few bespoke tailors, which once existed in dense clusters in the City and proximate districts. Within the City of Westminster, Soho has emerged as an important site of applied design and cultural production, including films, video games and software (Roodhouse 2008). But other districts which comprise interesting and instructive design sectors are located within the City Fringe, a territory roughly coterminous with London's traditional light manufacturing, engineering and distribution districts of the 19th and early 20th centuries (Figure 7.3). London's City Fringe thus encompasses the old industrial districts of Clerkenwell (Islington),

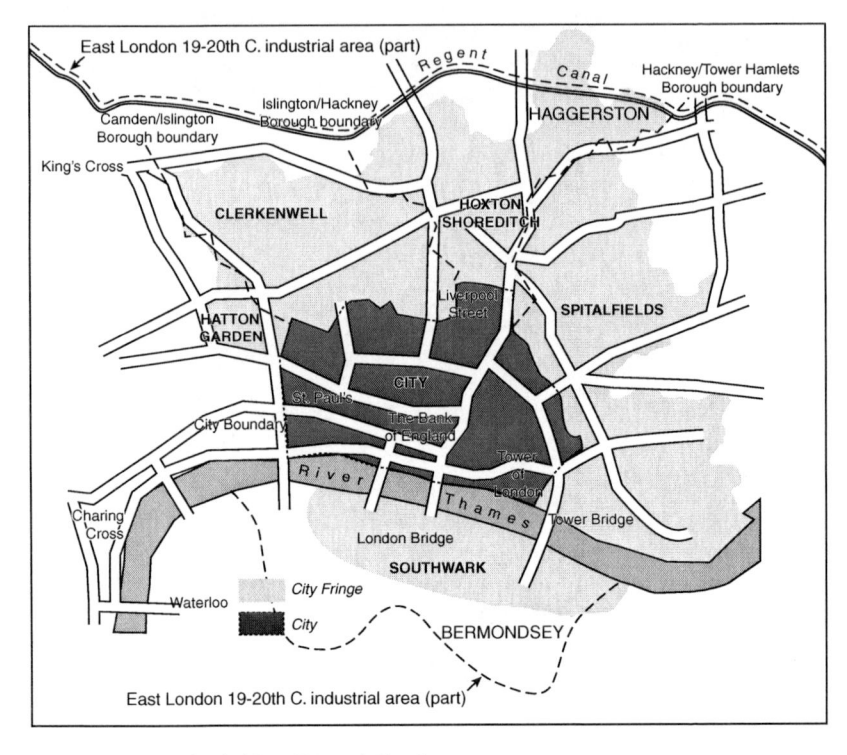

Figure 7.3 London's 'City Fringe' districts
Source: Hutton 2008

Shoreditch (Hackney), Hatton Hardens (Camden), and, south of the Thames, Bermondsey, in the London Borough of Southwark.

The City Fringe comprises an employment base of about 125,000, of which about 40,000 are employed within significant representations of design industries, including publishing, industrial, fashion and graphic design, architecture and landscape architecture, with one of the world's largest concentrations of new media industries and firms (City Fringe Partnership 2005). These design industries tend to occupy the heritage-built environment of London's heyday as a centre of light manufacturing and distribution, notably disused factories and warehouses, but also including former residential buildings, purpose-built studios, and office space.

The expansion of new media in the City Fringe is associated in large part with the talent and creative capital of the artist community of London's East End, perhaps the largest such aggregation in Europe, as well as with a layering of professional design firms within these landscapes. During the 1970s and 1980s artists infiltrated much of the old East London working class housing and proximate work spaces, with the pre-existing households unable to retain tenure in the face of income losses ensuing from the rapid collapse of traditional manufacturing and industry in the capital. Many of these artists maintained a kind of subsistence living, derived from intermittent sales of art and from transfers, such as grants and subsidies. But a significant minority of artists succeeded in commercializing their output, for example, from gallery exhibits or more informal sales, from contract work for professional design firms or other business clients or from multimedia work, for which London represented a major international centre of experimentation and innovation. Artists were crucial to the growth of new media in the 1990s, deployed in sketching, design, and drawing, combined with new production and communications technology which constituted, in the aggregate, much of the energy of the 'New Economy' (Hutton 2004). When the technology sector crashed in 2000–01, taking many of the dot.coms of the era with it, the true value of artists and designers as the seminal actors in the emergent cultural economy was vividly underscored.

Artists and design firms within the City Fringe, although certainly subject to the vicissitudes of competition and market fluctuations, benefit from a platform of substantial institutional supports, including universities and colleges, schools of fashion design, special foundations promoting excellence in design (for example, the Prince's Foundation, located on Charlotte Road in Shoreditch), and non-governmental organization (NGOs) and community-based organizations (CBOs), exemplified by the Clerkenwell Green Association, an agency which has supported traditional artisanal skills and craft production in this former core of London's precision trades sector. These institutional and programmatic supports provide a measure of resilience to firms operating in the City Fringe, in comparison with those in other cities where such supports may be lacking, and now include a place for design in local (borough) regeneration programs as well as those sponsored by the Greater London Authority and by central departments, notably the Department of Culture, Media and Sport.

The geography of design in London: An illustrative sketch

While design firms are found in many quarters of London's central and inner city, there is at some level a distinctive geography of design activity in the metropolitan core, following to an extent the tendency toward area-based specialization characteristic of the traditional East

End industries (see Martin 1964 in this regard). Some of these districts maintain long-established specializations, including jewellery design in Hatton Gardens, on the western zone of the City Fringe, bespoke tailors (although much diminished from the dense congeries of tailoring shops c. 1880–1960; see Martin 1964), and at least a residual of the precision crafts and trades of Clerkenwell, survivors of the brutal and comprehensive restructuring experiences of the 1970s and 1980s.

We can develop a profile of design industries in London's City Fringe that includes particularly significant representation in the districts of Clerkenwell (London Borough of Islington), Shoreditch (Hackney), and Bermondsey (Southwark). In the case of the former, there is a particularly rich presence of design industry firms, with important concentrations in multiple clusters of specialization (Figure 7.4). These include, notably, architects, graphic designers and related technical services, located in a tight cluster along Cowcross Street (just north of Smithfield Market), and a strip of architects, advertising and corporate branding firms and related services along St John Street, on the eastern margins of Clerkenwell. These two concentrations of design firms have enjoyed the benefits of proximity to the rich client base in the City of London just to the south and to transportation interchanges (including Farringdon Underground station), and a conducive ambience which includes the built environment, a rich amenity base and the historicity of Clerkenwell as a paradigmatic industrial district of London.

Clerkenwell's vocation as a centre of design industries is further underscored by other sites within the district, notably those of the Clerkenwell Workshops, situated in Clerkenwell Close. Indeed the recent history of the workshops discloses insights as to the larger transformations of Clerkenwell's development. Initially built as a London school warehouse (1895–97), the building was converted to artisanal workshops in 1975, with inexpensive space for over one hundred craft workers and designers. But a recent change of ownership has led to a refurbishment and rebranding of the Workshops as space for elite design firms and cultural industry enterprises, with the previous tenants making way for high-end architects, film production, corporate branding and new media. The strength of Clerkenwell's momentum toward high-end designers is further reinforced by the Business Centre property located at the end of Sekforde Street, a high-amenity conversion of a former workshop now dedicated to tier one design firms and cultural industries. These include music production firms, designers, interior design and corporate branding. As a representative example Bose·Collins has a portfolio which includes art work for recorded music production, as

The Geography of Design in the City 155

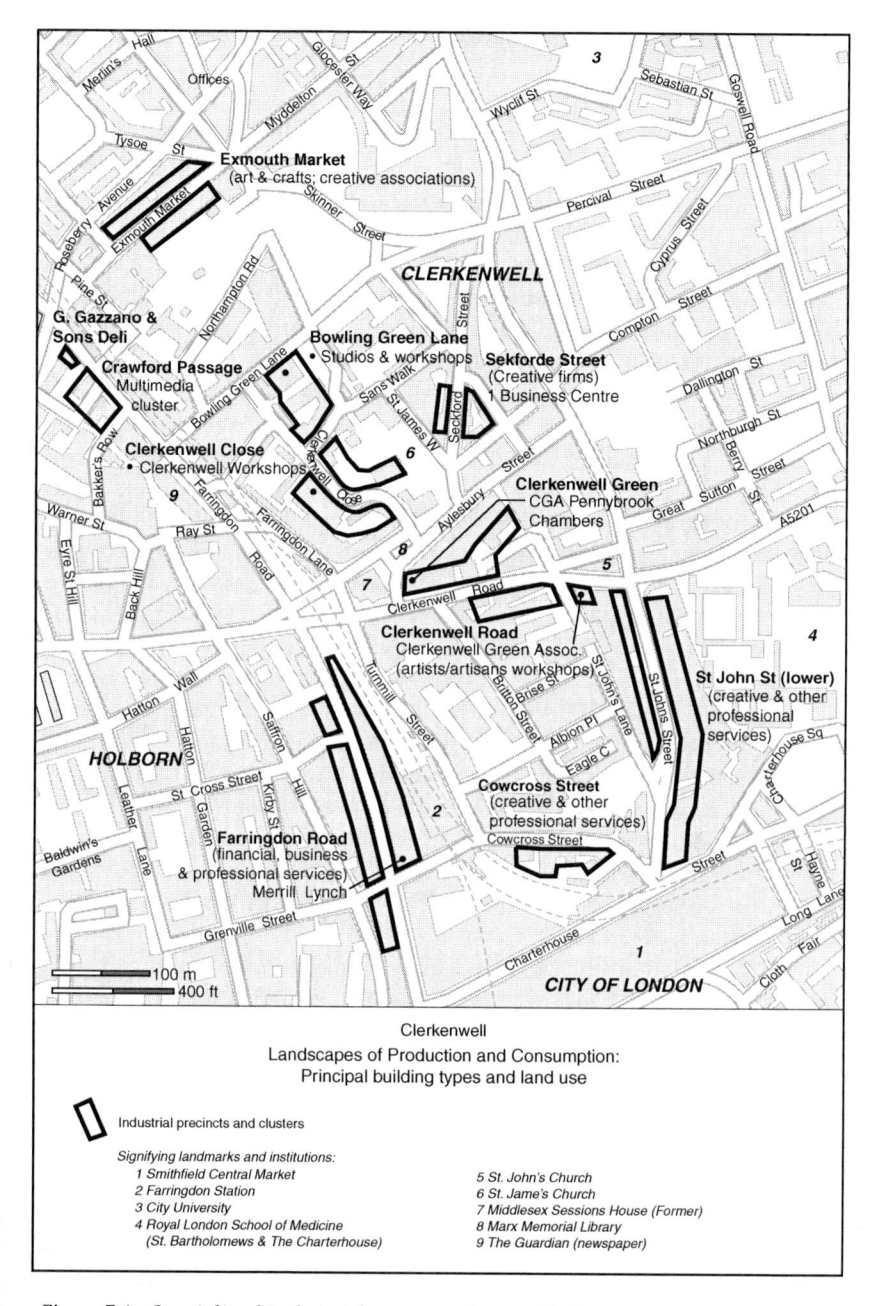

Figure 7.4 Specialized industrial zones and sites, Clerkenwell, London Borough of Islington

Source: Hutton 2006

well as consultancy work for the rebranding and relaunch of restaurants and bars in Chelsea, among other London-based projects.

If Clerkenwell's development trajectory exemplifies the shift from artisanal design and craft production to high-end professional design within a traditional inner city industrial district, Bermondsey offers (at a smaller scale) some instructive insights on the affinity between spatiality, aestheticized landscapes, and the geography of design in the metropolis. Bermondsey Street is situated just south of Bankside, and its centrepieces of London's cultural economy include the Tate Gallery of Modern Art, the Globe Theatre and the Design Museum in Shad Thames. Bermondsey's history as the centre of warehousing and distribution for products including spices and other foodstuffs and leather, and smaller role as light manufacturing site, has produced a legacy of high-integrity Victorian industrial buildings, the value of which has been recognized in the designation of the Bermondsey Street Conservation Area. The conservation of the area's distinctive built environment is supported by British Heritage, the London Borough of Southwark and other public organizations as well as NGOs and private interests, and is reflective of a larger public purpose in shaping a cultural economy vocation in this strategic but hitherto unfashionable borough of inner London south of the Thames.

The careful preservation and restoration of the industrial built environment of Bermondsey Street has attracted new cohorts, including creative industries, cultural institutions, artists and designers. Bermondsey Street and its proximate area thus incorporate an array of design firms including architects, interior designers, graphic artists and new media firms, together with a rich amenity base of restaurants, bars and coffee shops, replicating on a smaller scale the creative industry formations of districts north of the river (Figure 7.5).

But apart from these generic features of the contemporary urban cultural quarter, with its standard mix of specialized production and consumption activity, Bermondsey Street also boasts some larger enterprises which provide it with a larger projection within London's economy of creative industries. Perhaps first among these is Zandra Rhodes textile museum and salon, which occupies a prominent former warehouse in the heart of Bermondsey Street, its bold pink and orange external paintwork proclaiming its larger mission. The building's preservation and redesign was undertaken by Ricardo Legoretta, a prominent architect based in Mexico City, but with an extensive international practice, and local architect Alan Camp, who contributed local knowledge as well as the idea of developing residential units at the back

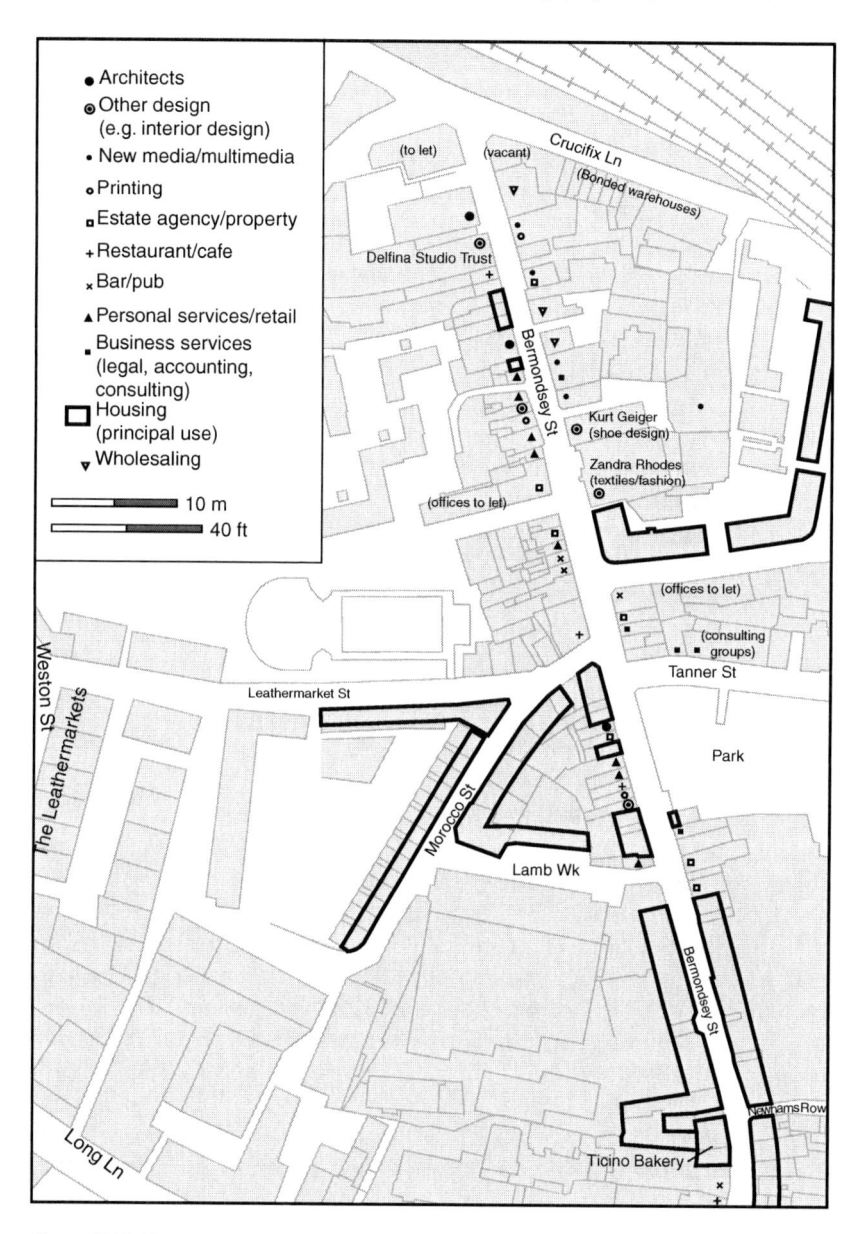

Figure 7.5 Design industries and firms, Bermondsey Street (Southwark)
Source: Hutton 2008

of the building which would cross-subsidize Zandra Rhodes' exhibition and salon space. Just up the street, the Delfina Trust operates an important program for young artists, both British and foreign, with attractive studios located in the upper floor of the building made available to winners of juried competitions sponsored by the Trust. At street level in the same building the Delfina Trust operates an upscale restaurant, providing a revenue stream for the studio space, and incidentally underscoring the associations between the arts, cultural production, and consumption in the contemporary economy. As a final local example, The Leathermarket, operated by the Workspace Group Plc as part its portfolio of managed workshops and studios in London, offers space principally to creative start-ups, with perhaps a sharper business edge than similar spaces operated by NGOs. Workspace Plc's mission entails both an overtly business model of entrepreneurship with a local economic regeneration purpose, demonstrating another facet of the design sector in the metropolis.

These vignettes of emergent design industries in selected London districts demonstrate a measure of variegation in terms of specific industry concentration and developmental storylines. That said, a thread running through each is the centrality of design to innovation and to creative production, and here the 'London labelling' aspect is crucial to the branding of cultural industries in the metropolis.

Design and new production spaces in Vancouver

Vancouver is roughly one-quarter of London's urban scale (by population), and lacks the world-scale concentrations of 'power sectors' (banking, finance, head offices, and political institutions) or propulsive industries which define the tier-one global city. But Vancouver has a well-developed cultural economy of creative industries, institutions and labour, with a number of specializations in design that underscore its value as a case study. These include, notably, architecture and urban design, video game production and new media: that is, industrial imprints of the sequences of industrial innovation and restructuring of the past two decades.

As is the case for many cities, there is a distinctive geography to the conceptualization and production of design in Vancouver, manifested by cluster formation and intricate network relations within the variegated districts of the metropolitan core, and with generally smaller (but dynamic) or incipient concentrations within suburban areas (Sacco 2007). To an extent this spatial tendency is generic, follows the widely-observed reassertion of specialized production within inner city districts among both advanced and transitional societies and includes concentrations of artists' studios and galleries in the East End of the City.

But there are some distinctive features of Vancouver's development which impart a measure of localized differentiation in the array of specialized design functions located within the metropolitan core. First, Vancouver's trajectory can perhaps more aptly be described as 'post-staples' than 'post-industrial', as the City never developed as a centre of Fordist production as did other, older Canadian cities in the heartland regions of southern Ontario and the St Lawrence (Barnes et al. 1992). Vancouver's formative inner city accommodated a substantial base of resource-processing and manufacturing industries, notably in the forest products sector, as well as foundries, food and beverage production, tanneries and the like. This (principally) staple-processing industrial sector was complemented by an extensive system of warehousing, distribution and associated infrastructure, including the western terminus of national rail systems, and by the development of Canada's largest port. In the aftermath of the collapse of the City's resource economy (accelerated to a degree by rezonings and land use conversions), the legacy of under-utilized space and abandoned industrial buildings provided the critical environment for the development of a new production economy in the core, which bears the imprints of each of the abbreviated restructuring episodes of the last quarter-century. In this reconfigured core zone space-economy, the CBD still encompasses the largest concentration of firms and employment in the City (and the metropolitan region as a whole), but now forms part of a more complex geography of specialized industries, labour and production (Figure 7.6). Amid the diversity of these new production spaces and sites in Vancouver's metropolitan core, the central role of design in new industry formation, labour practices and task specializations is underscored almost everywhere.

Secondly, Vancouver's post-staples development pathway has been accompanied by a 'post-corporate' syndrome, which takes the form of an attrition of head office activity since the 1980s, reflecting the concentration of corporate control in larger world cities ensuing from the rounds of mergers, acquisitions and takeovers which comprise part of the ensemble of globalization. As one consequence for Vancouver, parts of the CBD office complex have been stripped of corporate control operations, and have been steadily 'recolonized' by more contemporary industries, such as video game activity and new media. Thus the CBD has lost some of its pre-eminence as a mainstream business centre, and instead has been recast in part at least as a site of more 'diverse specializations', rather than a stand-alone agglomeration of business activity.

160   *Thomas A. Hutton*

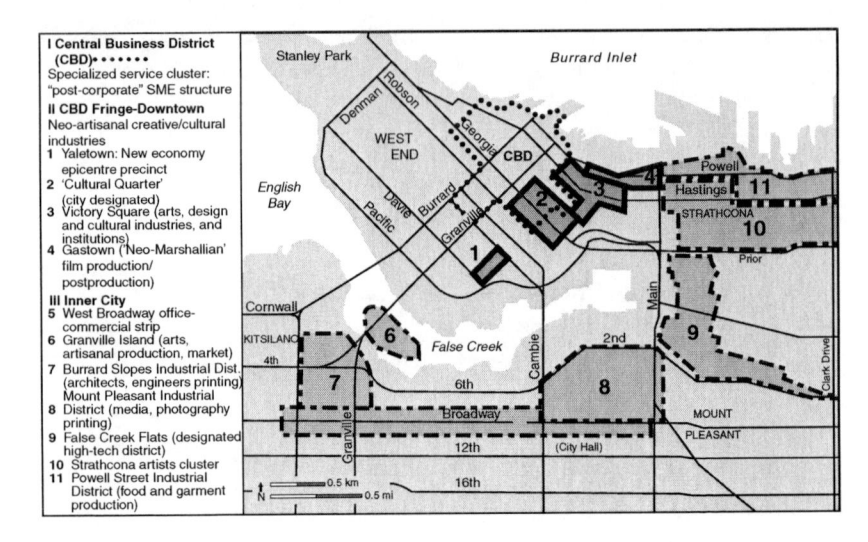

Figure 7.6   Specialized production areas in Vancouver's central area
Source: Hutton 2008

Thirdly, the decline of the City's staple economy, combined with the dramatic growth of international immigration since the 1980s in particular, has produced both a more multicultural, transnational urban identity, as well as an emergent economy tied increasingly to the markets, cities and societies of the Asia-Pacific. The cardinal impact of international immigration in the City's economy has been the establishment of a vibrant, entrepreneurial workforce which has supplanted the resource sector as the most influential force in the industrial restructuring of the last two decades, but another quite salient effect of transnationalism has been on the combination of cultural influences on design. While the durability of many of the classical centres of arts and design were derived from unique foundational cultures, Vancouver's cultural identity (and those of other heterogenetic cities) is increasingly shaped not only by the relayering of specific design and symbolic values, as new cohorts arrive in the City and join the workforce, but also by syntheses of multiple design traditions, expressed in the creative fusion of diverse arts, design, and symbols.

These broad processes of change in Vancouver's development trajectory, coupled with the sequence of industrial innovations over the past twenty years or so, have produced a palette of production sites which demonstrate the saliency of design to the contemporary urban economy. This specialization is most marked in the case of Yaletown (Figure

7.6, area 1), which has transitioned from a declining warehouse district on the periphery of the downtown, to the status of Vancouver's New Economy epicentre site. Yaletown encompasses major concentrations of design-intensive industries such as architecture, computer graphics and imaging and software design, all complemented by the City's richest amenity offerings. Yaletown also includes high-end loft conversions, contributing to the area's position at the apex of the City's rent structure. Victory Square (Figure 7.6, area 3), the old commercial and financial district of the City, bypassed when the centre of gravity for new investment moved west into the downtown peninsula in the 1970s, presents an altogether grittier ambience, which has served to attract start-ups in Vancouver's cultural economy, including many artists and professional designers. These enterprises are supported by a particularly rich platform of supportive institutions, including the Vancouver Film School, a leading international institute for video game skills training and a Centre for Contemporary Arts, to be located in the major Woodwards redevelopment project on the western margins of Victory Square. Just to the East, Gastown, the City's historic townsite, is the location for many architects and graphic designers, but its larger role is as an important locus of film production and postproduction – an important part of what Neil Coe has described as Vancouver's 'neo-Marshallian satellite production site' (Coe 2001).

Beyond the downtown peninsula, the inner city incorporates other sites of design-intensive industries, including a major cluster of architects and artists on Granville Island and the Burrard Slopes (Figure 7.6, areas 6 and 7 respectively), while numerous technical service operations in Mount Pleasant (area 8) comprise essential features of the design production sector of the core. The development of False Creek Flats (area 9), designated as a prospective high-tech New Economy site in 1999, just before the crash of the dot.coms, has been slower, but this relatively large district includes important design sites, notably Radical Entertainment, a major video game publisher, and the Great Northern Way Campus, which now offers a Masters degree in digital arts, supported by a consortium of four major area tertiary education institutions, and which aspires to offer education combining the attributes of arts, design and technology widely acknowledged as central to the formation of the cultural economy of the city.

We can conclude this succinct sketch of the geography of design in Vancouver with reference to some important dynamics of change. First, the profusion of architects which once clustered in the offices of the CBD has now largely relocated to the CBD fringe and inner city, underscoring that zone's high powers of attraction for design-intensive industries. Secondly, the transnational nature of this increasingly multicultural city has produced a fluid design culture, which is expressed in formal art, graphic design and architecture, among other fields. Young designers in particular are shaping a fusion of design values, symbols and influences which contributes to the City's emergent identity. Thirdly, although the tight production networks of primary design firms and a host of technical support service firms remains important elements of the geography of design, these are being increasingly stretched by insistent outsourcing, facilitated by constant innovation in telecommunications technology and systems. As an example, Radical Entertainment (see above), which develops games for Los Angeles publishers (and is owned by Vivendi, of France), contracts a portion of its drawing work to Chinese firms, as the quality approaches that of local artists and designers, and offers substantial cost savings in this competitive industry. Fourthly, while design is highly concentrated in the core, there are now important pockets of design capacity in the suburbs, including aerospace design subcontractors in the inner suburban municipality of Richmond, film production work in Burnaby and North Vancouver, and applied arts and design in Surrey and other areas. Finally, the tenure of design firms in the metropolitan core is in some cases compromised by the inflation in property values and rents driven by new housing, particularly in cities such as London and Vancouver. In these cities, and in others, the tension between the social reconstruction of the core and the sustainability of a cultural production sector has emerged as a major local planning and policy issue, as the contemporary redevelopment of the post-industrial terrains of the inner city continues apace.

Conclusion: Continuities and dynamism in the geography of design

In this chapter I have been concerned with developing a geography of design activity in the city, acknowledging the centrality of design to the cultural economy, creative industries, and constituent labour formation. A starting point for the exercise was a rehearsal of Allen Scott's model of industrial location in the metropolis, introduced in 1982, with a significant elaboration in 1988, and the elucidation of a spatial schematic of design activity that I presented earlier (Hutton 2000). A second important reference point concerns the developmental interdependencies between design, creative industries and specific cities as distinctive urban places which confer competitive advantage through signs and symbols, following the relational logics of production acknowledged in the Rusten et al. article on 'Places through products and product through places' (2007). The power of these interdependencies was powerfully observed in our vignettes of the London and Vancouver cases.

These conceptual models underscore the attraction of the metropolitan core for a host of contact-intensive design industries. Following the arguments of Marxists concerning the idea of a rent-gap, the early artists and designers were no doubt attracted to the inner city by low rents and the availability of space. But this account fails to offer a compelling explanation of location, and the differentiation of experiences from place to place, as a substantial body of research has generated insights into a much more extensive range of factors of attraction. The logics of location for design firms in the core include the distinctive spatiality and built environment of the inner city, the proximate supply of skilled labour, the rich amenity base of the central city, concentrations of clients and the proliferation of supportive institutions, agencies and programs in the core, relative to other zones of the metropolis (Bianchini and Ghilardi 2004).

To a large extent the geography of design activity has taken the form of industry clusters situated within the central and inner city, as exhibited in the London and Vancouver cases. But a more nuanced perspective discloses some important variants and departures from the classic clustering model, as in the engagement of design firms within more extended regional production networks, in subcontracting arrangements and in outsourcing. Over time these features of the geography of design are likely to increase in importance, following tendencies in other production regimes and sectors over the last three decades. Design will likely emerge as part of a 'new' new international division of labour (NIDL: Fröbel et al. 1980), in which less central places can become more prominent, abetted by the power of the Internet and other forms of telecommunications, and client demand, producing new production geographies.

That said, the inner city in many cases appears to possess a uniquely substantial platform of capital stock – economic, social, cultural, environmental and institutional – which should serve to anchor a substantial element of the cultural economy in place. This is especially true for those activities which encompass a mix of production, consumption and spectacle, or rely on proximate specialized labour and institutional supports. Apart from the well-known European and North American sites of design and cultural production, there is now a roster of important inner city sites to acknowledge, such as Ropponggi (Tokyo), Suzhou

Creek (Shanghai) and Telok Ayer (Singapore), among others, suggesting new possibilities for research on the geography of design. In these cases, as well as in the London and Vancouver vignettes presented in this chapter, the insistent process of property inflation and high-end residential development may at least in the short term impose greater dislocative pressures on the inner city's spaces of intensive design and cultural production than the allure of more distant territories and regions, constituting another dimension of the contestation between new industry formation and the social reconstruction of the city.

References

Barnes, T.J., Edgington, D., Denike, K. and McGee, T.G. 1992. 'Vancouver, the province, and the Pacific Rim'. In Wynn, G. and Oke, T. (eds), *Vancouver and its region*. 177–99. UBC Press, Vancouver.

Bathelt, H. and Boggs, J.S. 2003. 'Towards a reconceptualization of development paths: Is Leipzig's creative industries cluster continuation of or a rupture with the past?', *Economic Geography* 79(3), 265–93.

Becattini, G. 1990. 'The Marshallian industrial district as a socio-economic notion'. In Pyke, F. and Becattini, G. (eds) *Industrial districts and inter-firm cooperation in Italy*. 37–51. International Institute for Labour Studies, Geneva.

Bianchini, F. and Ghilardi, P. 2004. 'The culture of neighbourhoods: A European perspective'. In Bell, D. and Jayne, M. (eds) *City of quarters: Urban villages in the contemporary city*. 237–48. Ashgate, Aldershot.

Butler, T. and Lees, L. 2006. 'Super-gentrification in Barnsbury, London: globalisation and gentrifying global elites at the neighbourhood level', *Transactions of the Institute of British Geographers*, 31: 467–87.

City Fringe Partnership 2005. *Analysing the creative sector in the city fringe*. TBR Economics, London.

Coe, N. 2001. 'A hybrid agglomeration? The development of a Satellite-Marshallian district in Vancouver's film industry', *Urban Studies* 38, 1753–75.

Davies, L. and Hall, P. 1996. *Four World Cities: a Comparative Study of London, Paris, New York and Tokyo*. Bartlett School of Planning, University College. Comedia, London.]

Department for Culture, Media and Sport 2001. *Creative industries mapping document*. DCMS and the Creative Industries Task Force, London.Evans, G. 2003. 'Hard branding the cultural city: From Prado to Prada', *International Journal of Urban and Regional Research* 27, 417–40.

Elliott, P. 2005. 'Intrametropolitan Agglomerations of Producer Service Firms: The Case of Graphic Design Firms in Metropolitan Melbourne, 1981-2001'. Master of Planning and Design Thesis, University of Melbourne, Faculty of Architecture, Building and Planning.

Evans, G. 2004. 'Cultural industry quarters: From pre-industrial to post-industrial'. In Bell, D. and Jayne, M. (eds) *City of quarters: Urban villages in the contemporary city*. 71–92. Ashgate, Aldershot.

Florida, R. 2002. *The rise of the creative class: And how it's transforming work, leisure, community and everyday life*. Basic Books, New York, NY.
Fröbel, F., Heinrichs, J. and Kreye, O. 1980. *The new international division of labour*. Cambridge University Press, Cambridge.
Hall, P.G. 1998. *Cities in civilisation*. Weidenfeld & Nicolson, London.
Hamnett, C. and Whitelegg, A. 2007. 'From industrial to post industrial uses: The loft conversion market in London', Environment and Planning A, 39(1), 106–24.
Ho, K.C. 2007. *The neighbourhood in the new economy*, National University of Singapore, Department of Sociology.
Hutton, T.A. 2000. 'Reconstructed production landscapes in the postmodern city: Applied design and creative services in the Metropolitan core', *Urban Geography* 21, 285–317.
Hutton, T.A. 2004. 'The new economy of the inner city', *Cities* 21, 89–108.
Hutton, T.A. 2006. 'Spatiality, built form, and creative industry development in the inner city', *Environment and Planning A* 38, 1819–41.
Hutton, T.A. 2008. *The new economy of the inner city: Restructuring, regeneration, and dislocation in the 21st century Metropolis*. London and New York: Routledge Studies in Economic Geography.
Indergaard, M. 2004. *Silicon Alley: The Rise and Fall of a New Media District*, Routledge, New York.
Judt, T. 2005. *Postwar: a history of Europe since 1945*. Penguin, London.
Kaika, M. and Thielen, K. 2005. 'Form follows power: A genealogy of urban shrines', *City* 10, 59–69.
Martin, J.E. 1964. 'The industrial geography of greater London'. In Clayton, R. (ed.) *The geography of Greater London*. George Philip & Son Limited, London, 111–42. Norcliffe, G. and Eberts, D. 1999. 'The new Artisan and Metropolitan space: The computer animation industry in Toronto'. In Fontan, J.-M., Klein, J.-L. and Tremblay, D.-G. (eds) *Entre la Métropolitanisation et la Village Global : les scenes territoriales de la Conversion*. 215–32. Québec: Presses de l'Université du Québec.
Power, D. and Scott, A.J. (eds) 2004. *Cultural industries and the production of culture*. Routledge, London and New York, NY.
Pratt, A.C. (forthcoming). 'Urban regeneration: from the Arts "feel goods" factor to the cultural economy: A case study of Hoxton'. *Urban Studies*, London.
Roodhouse, S. 2008. *Game on? A report on the interactive leisure software subsector in London*. London: Creative Industries Observatory: University of the Arts, London.
Rusten, G., Bryson, J. and Aarflot, U. 2007. 'Places through products and product through places: Industrial design and spatial symbols as a source of competitiveness', *Norwegian Journal of Geography* 61, 133–44.
Sacco, P.-L., with Del Bianco, E. and Williams, R. 2007. *The power of the arts in Vancouver: Creating a great city*. Vancity Capital, Vancouver.
Sacco, P. and Tavano Blessi, G. 2009. 'The Social Viability of Culture-led Urban Transformation Processes: Evidence from the Bicocca District, Milan', Urban Studies, 46(5–6), 1115–35
Scott, A.J. 1982. 'Locational patterns and dynamics of industrial activity in the modern Metropolis: A review essay', *Urban Studies* 19, 111–42.

Scott, A.J. 1988. *Metropolis: From division of labor to urban form*. University of California Press, Berkeley, CA.

Scott, A.J. 1997. 'The cultural economy of the city', *International Journal of Urban and Regional Research* 21, 323–39.

Scott, A.J. 2007. 'Exploring the creative city paradigm', presentation to the *Biennial Meeting of the Pacific Regional Science Association (Vancouver, 6 May 2007)*.

Storper, M. and Salais, R. 1997. *Worlds of production: The action frameworks of the economy*. Harvard University Press, Cambridge, MA and London.

Zukin, S. 1989. *Loft living: Culture and capital in urban change*. Rutgers University Press, New Brunswick, NJ.

Zukin, S. 1998. 'Urban lifestyles: Diversity and standardisation in spaces of consumption', *Urban Studies* 35, 825–39.

II
Design and Firm Competitiveness

8
Design and Gender with a Competitive Edge

Lena Hansson, Magnus Mörck and Magdalena Petersson McIntyre

Introduction

Firms compete on price and quality, but can also compete on issues of inclusion and sustainability, including gender and design as a competitive strategy. To develop a company brand that projects an image of an inclusive organization is one strategy that businesses can deploy to develop a unique selling proposition. Another possibility is to adapt product designs to include more groups within a population. This brings a new role for gender awareness in the construction of a design-based competitive strategy and situates gender in a unique position. Gender is hence seen as profitable in several ways. A substantial reason for businesses to make gender part of their marketing is also an awareness of the growth of female consumers and the importance of women in consumer choice.

We will analyze design in terms of gender theory and the possible implications for particular consumer segments; the market discourse that has developed in the U.S. will also be woven into our argument. Finally we conclude with a tentative conceptualization of our findings. Due to the complexity of gender, we will try to identify different aspects and also acknowledge that gender can mean paying attention to women's heritage, as well as men's. Everything about gender tends to be questioned. Take the example of a design activity like knitting; today it is highly popular, but it can also be interpreted in ways that are contradictory: to some a way to laud women's history, to others a bag of inventive and provocative textile graffiti and to yet others a future in which men are included in this activity. Gender theory touches upon sensitive issues – sometimes it implicates the idea of profound differences between men and women, but more often than not it moves in more ambivalent directions. We have interviewed female car designers who argue for positions that are a lot more delicate. These female designers who worked in industry were few and dependent, and hesitant to challenge masculine prerogatives in the definition of 'car culture'. They preferred to shift to issues of marketing, as a way of sporting their own special knowledge about female consumers. Their knowledge about gender was cast in terms of markets, as one way of avoiding criticisms for highlighting issues surrounding gender in the design process; discussion of gender issues easily disturbed conservative managers.

While the benefits of equal representations for organizations can be explained in fairly straightforward ways, the implications of taking gender into consideration during design processes is more complex and filled with contradictions. It involves a consciousness of the gender of the designers, the ways in which particular designs in themselves are gendered, as well as recognition of consumption patterns being gendered.

Gender tends to stand for women's rights, but men also have suppressed needs due to outmoded gender discourses and practices. When gender focuses on heterosexual couples there is a danger that same sex couples are neglected. In design terms, gender tends to fade into a simple dichotomy of forms, tastes and of clear marks of either/or male/female (Sparke 1995). The logic is to develop differences, instead of cultivating a more inclusive and moving design field. Here we argue for poststructuralism as a way of not predefining the heterosexual position (Butler 1990, 1993, Laclau and Mouffe 1985). In our analysis of the car industry we found men and women taking up a number of positions on gender and design that created an inventive bag of ideas and that tried to avoid making design always either male or female. Aligning poststructuralism and design means that the latter is considered as a subject position. Any individual is seen as moving between several positions, negotiating and shifting. If understood this way, females consuming male design and vice versa, our understanding of gender becomes more open to change and design intervention.

Examples that encouraged us to become interested in this issue include Dove's Evolution, a product marketed with the images of the human body that are far removed from the ordinary stereotypes used in advertising. Here companies can become interesting and add value to their brands. But the implications of gender and design of course go far beyond producing new imagery but also involve the development of goods that address issues in the areas of health, transportation, working clothes, home appliances and much more. As competition between firms increases due to globalization, it becomes more important to connect brands with specific social and cultural values. Thus, some companies develop associations between the concept of an equal society and their brand identities. In this chapter, we will explore how gender and design can be made to interact to enhance profitability and customer use values, as well as to create designer work of considerable interest that contributes to a better society in terms of gender-informed design. This chapter summarizes the findings of a two-year research project on gender, design and business organization. This was a multidisciplinary project which combined expertise from culture studies, marketing and design; the project was funded by the Swedish innovation agency, Vinnova, and some private companies. The aim of the chapter is to evaluate this body of knowledge and to apply the issue of design to improving the daily life of consumers. This involves identifying exemplar products on the market and products that we sampled in cooperation with a design school. The task was to reflect on the implications of the market in influencing the result with the goal of identifying both the advantages and disadvantages of combining markets, gender and design work. Different approaches to gender articulated by designers and companies will be of special interest, since this is a volatile and controversial field.

The layout of our argument

The chapter begins with an analysis and brief discussion of the background of gender theory. We explore two strands: perspectives on gender as found in the field combined with a poststructuralist frame. In three sections we argue our case and present an analysis of our understanding of developments in the fields of marketing, gender and design. The first section, *Commercializing gender*, provides a survey of the relationship between the market and feminism. Here we try to clear gender and diversity out of the traditional blind faith in the market and try to remove the unprofitable misunderstandings that exist between the literatures of business and feminism. American marketers have developed different brands of market feminism that provide different understandings. We use poststructuralist theory to analyze a number of highly influential contributions that have been made in the field of gender discourse (Barletta 2004; Popcorn 2000; Scott 2005).

In section two the analysis shifts towards design issues, inspired by conceptual design, rather than industrial design. We survey a variety of consumer goods with gender implications and develop a typology to account for some of the variety that we have observed. This way we develop an inclusive position that has the merit of preserving some of the variety in the field and does not just align the analysis with queer theory or approaches to women in history that tend to aspire to hegemony. Our own brand of gender theory is one step removed: prepared to see change, ambivalence and diversity, while paying attention also to masculine, gay and lesbian markets.

In the market, the industry and the studios of the design schools we found design and gender relating to health, education, entertainment, transportation and much more. The third section will develop the case of the concept car Volvo YCC. This is the primary focus of this chapter. The concept car project was initiated after the visit to the company by one of the most important feminists in the field of marketing, Martha Barletta. When the car finally reached its apex of fame, it was lauded by Penny Sparke and written by her into the canon of gender theory, and this has created certain expectations. But not least important were the reactions inside the design team at Volvo which went through a profound learning process in experiencing the practice of gender and design as a source of success, but later sliding into defeat due to animosities that developed within rival factions of the management team.

Finally we offer our conclusions; we argue the possibilities of good business and use values in terms of gender and diversity, a field of controversy that does not invite the fainthearted, an insight that is difficult to deny.

Commercializing gender

Our interest in the 'commercial' has two objectives. One is to discuss the implications for private businesses of making gender visible and, in some cases, taking critical stances toward conventional identity positions such as gender as a means for enhancing competitiveness. The other objective is to challenge critical interpretations of 'the commercial' that have often figured in gender and cultural studies. Our starting point is that the cultural and the economic should be perceived as intertwined and mutually constitutive, a point that has been made by scholars of consumption and material culture (Jackson et al. 2000; Du Gay and Pryke 2002)[1]. Divisions between the cultural and the economic have been criticized by these scholars, and we believe that it is necessary to consider the ways in which these spheres intersect (Brembeck et al. 2007). Cultures of production such as design processes and entrepreneurship have in the last decade been more and more included in the field of consumption, particularly fashion studies that, according to the anthropologist Kaori O'Connor (2007), earlier tended to demonise capitalism. This is also in line with our own aim to dissolve distinctions between design, consumption and market strategies.

Within the last 10–15 years there has been an enormous increase in books and consultants emphasizing the importance of diversity politics for organizations that want to prosper in the future. There is, at least in Sweden, a well-known academic and feminist critique of these commercial attempts to transform equal representation and diversity policies into profit. Diversity strategies hide historically grounded power relations that are still operative and present them as assets for the commercial field (de los Reyes and Martinsson 2005). The essence of the criticism has to do with the ways that the participation of so called 'other groups' of the population, such as women, homosexuals and ethnic minorities are not invited in without any conditions attached. They are expected to bring exciting, different perspectives with them into work places. This is seen as reaffirming definitions of normality and difference. They are also expected to contribute along expected lines, regardless of individual preferences (Martinsson 2006). While we agree that these problems are inherent in many commercial diversity and gender strategies, and are important to analyze, our aim is to focus on the commercial possibilities that these strategies bring. We argue that over-looking the possibly positive effects of how identity positions such as gender operate in different ways in commercial strategies creates tendencies where markets are perceived as natural forces that cannot be altered and where all attempts to challenge identity constructions through commercial practise end up confirming the same identities. We believe that these commercial gender and diversity strategies are also bound to bring with them many changes in the ways in which different groups of the population are addressed through design, marketing and advertising. The motivations behind thinking of equal representation may be commercial, but the possibilities might result in new ways of performing one's gender.

As we will show in the case of Volvo YCC, describing gender as something profitable can also work as a strategy to introduce equality in firms where criticism of conventional gender roles is generally thought of as something negative or at least undesirable. We would like to clarify that 'gender' can have many different meanings in this context. Our own definition is social constructionist. Gender, like all identities, is an ongoing cultural construction and has (in contrast to many popular beliefs) no natural or biological ground that determines how it should be interpreted, acted or organized. Rather it is formed by and through social and cultural processes that include all human behaviour. It should be seen as a relation of power between people that we all partake in to make sense of ourselves, others, our bodies and our places in the social order. Also, gender is historically, socially and culturally variable. There are no known universal definitions of what gender is or should be which means that gender can always be performed differently than the way it is in a particular context.

More specifically we draw on Judith Butler's theories of the performativity of gender identities. Gender is enacted performatively, which means that it is not expressive of an inner essence and is thereby not prior to the various acts, postures and gestures by which it is dramatized and known; rather these attributes effectively constitute the identity they are said to express or reveal (Butler 1990: 278). These dramatizations of gender can, we argue, be extended to include design objects. Soft, rounded shapes are often associated with the feminine, while straight and sharp shapes have come to be coded as more masculine. Likewise, pastel colours are mostly considered feminine and men are supposed to prefer dark and natural colours. Such gendering of shapes and colours can be observed in a wide range of consumer products from baby fashions to cars, and from kitchen appliances to power tools. Making the point of the performativity of objects, we argue that these objects are, just as the acts, postures and gestures are, productive of gender identities, part of the performativity of gender. The clothes we wear, the cars we drive, and the tools we use are all producers of cultural fictions of particular identities. It is, however, important to point out that meaning-making is an ongoing process and there are no inherent values in objects of any kind. Angela Partington (1997) has described design as a discourse, rather than as a language, wherein it is not the objects in themselves that are interesting but the practises, specifically consumption, that give these objects meaning. This is a point that we also would like to make and from these perspectives we will discuss the essentialism inherent in the idea of marketing to women.

The commercial applications of 'gender' that we have encountered cover a broad spectrum of possible definitions. Some designers and market strategists that we met during our fieldwork shared our own ideas of what gender is – that is, a social construction. Others had not put a lot of thought into it and relied on common sense ideas of equality as something worth striving for. Others yet believed in biological differences between men and women that could not be overcome socially, and some, relying on more natural scientific approaches, worked on making physical differences between men and women visible without making any particular value judgments of the more social aspects of gender. An example of the latter is the development of a crash test dummy based on an average female figure (described below). The engineer/scientist behind this project was motivated to produce this dummy through car crash and injury statistics that highlighted a societal need for improving car safety. Gender was thereby reduced to statistical physical variabilities.

There have been some attempts to challenge the often negative depictions of markets that point to the contextuality of consumption practises. Within the field of marketing, feminist scholars such as Catterall et al. (2000) have pointed to the difficulties and potentially explosive combination of marketing and feminism. Marketing represents a system that traditionally has been heavily criticized by feminists for exploiting women and this has meant that it has been difficult for advertisers to be perceived as radical (Catterall et al. 2000). A related attempt to overcome the ways in which 'the market' is seen as a problematic, negative force in critical, feminist studies is found in the work of Linda Scott (2005). She claims that feminist traditions of criticizing capitalism and consumption have resulted in simplified polarizations of the market, on the one hand, and equality work, on the other. Many women have, however, through history reached emancipated goals through commercial activities, not the least within fashion, design and advertising. In an attempt to form a postmodern market feminism, the economist Deidre McCloskey (2000) claims that the market society has had the greatest potential for changing the position of oppressed groups as they struggles for liberation. Both Scott and McCloskey have been criticized for glorifying capitalism. The postcolonial critic, Gayatri Spivak (2000), has for instance objected to McCloskey by stating that there can be no natural link between market societies and liberation. Rather, while some will be liberated by capitalism, others are likely to be more intensively exploited.

Speaking of and doing 'the market' makes the market appear to be a natural given. Instead it should be analyzed as something that is created through specific power relations and that has particular effects. We want to emphasize the economic as something constantly in the making and as produced by and a producer of power relations and identity positions.

Don't think pink

Another way of addressing gender and design is through the well known strategy that can be labelled as 'don't think pink'. This is the essence of consultancy handbooks aimed at industries that try to develop a strategy for capturing female consumers. Martha Barletta (2004) and Johnson and Learned (2004) have written popular books on the subject (an earlier version is Faith Popcorn 2001). These books are based on the idea that men and women are *different*, biologically and socially. Women consumers are generally overlooked by marketing strategists and in order to capture their buying power it is necessary to speak to them in a respectful manner. Most of all women hate it when companies produce a smaller, cheaper version of something that is already out in the market, paint it pink and re-target it at women consumers, hence the expression 'don't paint it pink'. Or in Barletta's words: 'Refusing to acknowledge women buyers' different preferences and priorities won't make them go away – the preferences, that is. The women probably will' (Barletta 2004: xx).

From a gender analytical perspective these books are highly problematic. Many of them focus on demonstrating that men and women are essentially different. It is perhaps needless to point out that such essentialist claims are incompatible with an understanding of gender as socially constructed. The books cite researchers who claim that differences between men's and women's brains account for the social and cultural enactments of gender. Such claims are dangerous in that they legitimize social orders and inequalities by reference to 'nature'. These dangers should not be underestimated, and furthermore these books are wide-spread and respected within many commercial fields so there is every reason to take them seriously. It is also important to note that these books take for granted that heterosexual, middle-class, nuclear families are fundamental.

From the perspective of the economy and market described above, we, however, also want to examine whether the endeavours presented by Barletta (2004), Johnson and Learned (2004) and Popcorn (2001) can be understood as expressions of changes in the ways that 'the market' or consumer society interrelates with the construction of particular identities and subject positions. An interesting question to pose is also *why* the argument is built up around physiological differences and *whether* it would be possible to design strategies for marketing that challenge the assumed notions of men and women *without* falling into the pitfall of essentialism. The books are built up around tensions of, on the one hand, the importance of acknowledging differences between men and women and, on the other hand, of showing that women and men want the same things; not all women want pink products and they want products that provide the same technological performance as those men want. Then what is the point of emphasizing differences based on biology and neurology, one might wonder? While asserting biological differences, these books simultaneously dissolve normative assumed notions of gendered design and marketing ideas by bringing invisible norms into the open. They question the order of things and cause trouble by questioning the male as an invisible norm for all humans and by questioning stereotypes such as that women like pastel cars with low performance engines.

Market segmentation

Since the 1950s it has become more and more common to develop niche products that are targeted at particular groups such as men, women or young people. Marketers no longer believe that the same product could appeal to people regardless of lifestyle (Sparke 2003). This development is generally referred to as 'market segmentation' and accounts for many of the gendered products we see on the market today. Market segmentation has become marketers' and advertisers' common approach for increasing sales and market share. Even though manufacturers of, for example, toy soldiers and dolls always had a girl or a boy in mind as the user of the products – not just a child – these processes have been speeded up today and include a wider range of products. In today's toy shops there are virtually no objects that do not have either a boy or a girl as their intended user. These market segmentation processes not only tell us that boys and girls, and men and women like different products, because they are different. They also tell us that either you like girl's toys or boy's toys. That's it; there is nothing in between or to the side. There are only two possible gendered subject positions to choose from, you are either one or the other, not both (Loxley 2007).

Gendered design practice

In this section we try to grasp some of the complexities that come from juxtapositioning gender and design by presenting evidence of different marketing strategies. These strategies will be illustrated by examples from a number of different industries. Design has the potential to make the world a better place by providing consumers with objects and environments that make not only their everyday life easier and more comfortable but also more beautiful. Reflecting societal and cultural values as well as commercial and political interests of its time (Thackara 1988; Woodham 1997), design is an important tool for creating commercially successful consumer goods but it also influences these values and can be blamed for contributing to reproducing gender in old and outdated fashions. This preserving role of design is by no means definite; on the contrary, it implies that design can lead the way for societal and cultural change by influencing consumers' views and attitudes towards how things should be and how they should look. Design thus also makes moral and political statements and in terms of equality, gender has been one of the concepts at work. Few projects apply equality and gender to design and commerce, which makes our project quite unique.

Design has been used for political ends before, as the democratic ambitions of Bauhaus or the arts and crafts movement witness. A democratic idealism has been a basic premise of Scandinavian design since the beginning of the last century (Sparke 1987). A more modern or recent design movement that promotes equality and diversity is Universal design (Ostroff 2001). Although it tends to focus on age and ability in practice (Hansson 2006), gender is another important issue embraced by this design philosophy, which opposes categorizations of people and discrimination against people who do not fit the norm (Welch 1995). Universal Design should not be interpreted as an impossible standard or as a 'one size fits all' approach but as a consumer market-driven issue and a quest for more alternatives. Individual consumers can then choose in accordance to their preferences without becoming stigmatized or excluded. If design excludes it becomes a social and cultural issue, whether it is because of neglected physical needs, for instance, consumers with too small hands who are unable to manage the tools of their trade, or cultural needs, that is, not enough alternatives to make an aesthetically satisfying choice that enhances consumers' need of self-expression.

Interestingly enough, objects are commonly viewed as neutral and when reflected upon, it is usually design aspects like function, expression and style that are considered, not gender (Jahnke 2006). As part of the project we scanned the market for good examples of how gender issues can be approached and communicated through designed objects. In our analysis we found six strategies that in one way or another challenge traditional or gender-ignorant design (Jahnke 2006).

The first strategy is *moving the gender dichotomy by mixing*. Unisex was the solution for bridging the gender gap during the 1960s, part of the youth revolution when young people also became prime actors in consumption. Young bodies, men and women, modelled similar fashions (Lurie 1983). Today it is a growing trend within children's clothing and toys in Sweden. These markets have become one of the most stereotyped in terms of gender which is obvious when you walk into any retail store or toy store, where the girls' section is separated from the boys' both in terms of different gender coding of clothing and toys and the display. Unisex then provides an alternative to this categorization of girls and boys and works as a gender-reducing strategy. The problem is that in an effort to make gender neutral objects, designers may choose to use genderless colours and forms and this might produce rather dull design solutions. Going for a gender-reduced strategy or a 'neutral' design, designers must also be aware of how, in the design process, the male norm is taken for granted and may appear as the natural and 'neutral' option. However, unisex also survives and prospers without that name in sport clothing; different cutting is required for clothing to fit the female or male body, and sizes vary, but it is not as stereotyped as the coding scheme is more or less the same and the focus is on fitness and training as activity rather than superficial gender coding. It could be summarized as the same but different. But in the stores categorization is used for easy access.

The second strategy is *forgotten needs of the female and male body*, an important source of innovation and good business. Focusing on similarities between bodies, important needs might be neglected. Here we have working clothes as an example. When women entered industry they usually had to and still often have to wear clothing that has been designed and made for men. Wearing clothes made for men not only makes women feel uncomfortable, as they are impractical, but also causes emotional discomfort because they are ill-fitting and coded in masculine ways. Adapting these clothes to women might not be the best option; a more optimal solution is to start from scratch, working with a female model. Adjusting a male model to a more feminine one usually only involves superficial changes, for example, including a seam for a more bottle-shaped body. SVID (Swedish Industrial Design Foundation) and the Swedish Trade Union Confederation (LO) have together with SSAB, a Swedish steel manufacturer, developed a working overall for women based on the average female body.

Design based on the man as norm is probably the most used strategy on the market, resulting in the neglect of women's needs. Sometimes this neglect might have very serious consequences, for instance although the number of female drivers is about the same as male drivers studies have revealed higher levels of whiplash injuries amongst female drivers caused by seats that have been designed for the male body. The crash-dolls used to test cars were male and the first female crash-doll based on an average woman was only developed in Sweden in 2006. Earlier the standard test doll was male and at most was modified for a short or sometimes pregnant woman.

The growing market for grooming products for men implies that the market had previously neglected men's needs and demand for skin care products and cosmetics. New product lines for men have been developed by brands like Biotherm and Calvin Klein and in department stores special sections are designed to appeal to what is believed to be the masculine taste. The market works here as a vehicle of change making it more acceptable for men to participate in the grooming industry. However, the market is also responding to the emergent image of the 'metro man', the 'new English dandy' or the 'new man' (Cicolini 2005, Nixon 1996), a modern heterosexual man who is interested in grooming, style and fashion, something that does not make him feel any less masculine. In addition, this need for men to express themselves in new ways is a border crossing; they are stepping into a traditional female domain, something that will be further discussed in strategy five.

Supplying the market with designed objects that exploit neglected needs is taken further in the third strategy, *beyond the gendered body*, which embraces diversity in needs without defining them as male or female. It opens up for inclusion into the market more people but not all people. Small hands are a characteristic of some women and men. Catering to this group involves, for example, providing grips for small hands and these products include everyone other than those with large hands (mostly men). A German company, RTI Sports, has marked a new kind of bike handle, Ergon Grips. They employed the world famous Swedish design company Ergonomidesign to design these ergonomic handles that can be adjusted for different sizes of hands; smaller for women than for men, and with different amounts of cushion. The aesthetic expression is the same and the smaller version could easily fit a man with smaller hands as well, but then RTI Sports wanted to target women specifically in their advertising by making the packages with different colour schemes. Here the focus on gender is replaced by bodily size. The point is by suppressing gender as a focus, the product will be as accessible to women and men.

Women crossing the border is the name of the *fourth* strategy. The brand Barbara K has acknowledged the enhanced importance of do-it-yourself (D.I.Y.) and provides women with power tools. The electrical screwdriver is lighter than usual as the battery is removed from it, something that could benefit anyone. But their business strategy is to address women's needs, which the less macho expression of the electrical screwdriver also implies. Using codes that open up and soften heavy masculine design is one way of making power tools accessible to women. Although a gender perspective tends to focus on how women's needs are neglected in the design of objects, there is also evidence of the contrary, neglect of men's needs, especially in domains traditionally defined as female.

There is thus an equivalent *fifth* strategy to the previous one, *men crossing the borders*. With the increased equality between men and women, the male role as a father figure has changed. More men are on parental leave and the market has responded; nowadays you find a range of consumer objects that address men who take care of babies and childcare, like diaper bags, baby carriages and baby carriers. As the male norm also restricts men's behaviour, not only functional needs are covered but also men's needs to express themselves. That is why these objects are made masculine or perhaps more correctly, less feminine. Pippilotta, a Swedish-based web store for families with children , market their 'Diaper Dude-camo' as 'a sporty and luscious diaper bag for cool fathers (and mothers)' (http://www.pippilotta.se). Although it is primarily marketed at men, they do include women in their communication as a camouflage coloured object may be attractive for women as well. Pippilotta also markets a female counterpart, the 'Diaper Diva-Pink Camo' but this time the invitation to men is obviously seen as more of a risk, as it is ended with a question mark followed by an exclamation mark '...to all mothers (and fathers?!)'.

Last we have strategy six. *Breaking the mould of heteronormativity* means, for both men and women, that they question the basic dichotomy of having gender differences organized around biology and heterosexuality. This area brings forth queer design, a culture of design excess with influence on fashion and home decoration. Queer design belongs to postmodernism, and has paved the way for design movements like Memphis. It is characterized by glamour but also humour and gender-disrupting objects.

As we have illustrated earlier the market can be an actor of influence in changing the borders of what is acceptable and not in terms of performing gender. The workwear kilt designed for Blåkläder, a manufacturer of working clothes (Figure 8.1), will be used to exemplify how this can be achieved and is another of our case studies.

It started out as an idea in the head of the designer Marcus Jahnke, combining functional needs and a gender thought. Men have a much more limited coding scheme for clothes that restricts their possibilities to perform gender differently. Interestingly enough, this kilt made its way into a very macho domain, the construction industry, and became a market success, despite its somewhat provocative gender statement.

Figure 8.1 Workwear Kilt, Blåkläder; designer: Marcus Jahnke; photo: AORTA
Source: Courtesy of Blåkläder

At least in a Swedish cultural context, there are still relatively few men who would consider wearing such garment, even though the term 'kilt' is associated with the traditional Scottish garment, worn by men. The workwear kilt might as well be associated with the skirt, a typically female garment. But a masculine coding can be enhanced by advertising. In the advertisement, the kilt is worn by a man standing in a very masculine posture – his arms are crossed and he maintains a solid stance, legs apart, gazing into the camera. The kilt looks masculine partly because of its wearer and the milieu created for the picture, but also with the help of design elements, the use of fabric, thick seams and pockets for tools. This helps to legitimize the otherwise somewhat contradictory or provocative clothing amongst building workers even if it still is a gender-disrupting materialized sign. It has gained international acknowledgement and was awarded the best fashion product of the year award (2007) by the Swedish branch organization of the textile and fashion industry. Today, the work-wear kilt is part of Blåkläder's regular collection because of the positive response that the company received from its customers.

These six strategies are inclusive if not 'universal' because they consider consumer needs that have not been recognized before. Some of the identified strategies only emphasize physical needs but several include an aesthetic need for self-expression as well. However, the context cannot be dismissed in the analysis because what is an accepted norm in one industry might be very different from how things are done in another and one has to take that into consideration. In a very male dominant industry, just considering women's needs might be a great step forward although its success may differ depending on what strategy is deployed. One can never be sure of the response from the market. Even if women's needs are addressed a product will not appeal to all women, and although it might very well be an object that both men and women can use the marketing strategy will still influence consumer choice. The awareness of gender issues in design must be followed by the same awareness in marketing and retailing. An awareness of gender in design can, with the presentation of good examples, give rise to innovation, new design solutions that address new consumers and thus advance commercialized equality in the market. The difference might in some cases be almost invisible but the idea behind any designed object must guide the process, and with gender in mind, the end result can make a difference in the quest for equality.

Conceptual design

Our study also includes the acknowledgement of *conceptual design* and its ability to draw attention to certain issues, gender being one. Conceptual design has cognitive shock value and its dynamics have been researched in recently developed artistic research (Hannula 2005). It gains inspiration from conceptual art, claiming it is the question behind an object and the design process rather than the object itself that is important (Robach 2005). It is part of the critical design field, and its purpose is that with the help of objects, it is possible to stimulate a debate between the designer, the public and the industry (Emilson 2005). From a critical design perspective most design reinforces the dominant social, cultural, technical and economic values of a society

(Dunne and Raby 2001). There are far fewer critical designs that provide alternative perspectives to mainstream design.

A good illustration of how products are gender-coded or coded differently within different domains is Karin Ehrnberger's work as part of her exam work at Konstfack, Stockholm's art and design school. The issue is what is revealed if you alter the expression of a power drill and combine it with that of a hand food mixer – how would such a product appear? Think of a hand food mixer, used in the kitchen, traditionally a female domain. It is often white with a smoother form than a power drill, which makes it look rather innocent. But with its sharp blades it is far from an innocent tool. A traditional power drill on the other hand is green or red, a dark colour with black details. It looks powerful with a rather sharp form and the product references are to a traditionally male domain, the construction industry. Now, consider the redesigned drill illustrated in Figure 8.2.

Karin Ehrnberger shows the possibilities of *border crossing* by changing the design expression amongst these two very differently coded products. The drill suddenly looks easier to use and more harmless. In reality, both tools are rather easy to use but can also become dangerous if used carelessly.

Design elements like form, colour, ornaments, texture, material, graphics and so on are used to communicate qualities like function, ease of use, and product quality, not only aesthetics (Bloch 1995; Creusen and Schoormans 2005), to create an intended expression. But how these expressions influence perceptions of gender is often forgotten. These expressions also become prototypical or stereotypical over time and thus harder to change. However, consumers' yearning for novelty opens up possibilities for alterations in the appearance of products.

By creating models or prototypes of objects that help to perform gender differently it is possible to subvert routine thinking about how things should look that assigns men and women to segregated roles. As part of our project, artistic research has been undertaken using conceptual design as a means to open up awareness of how gender is considered and coded in designed objects. Students from the Swedish School of Art and Design (HKD) in Göteborg were tutored by designer Marcus Jahnke, and gender considerations were explored in relation to design and experimented with in the gestalt of objects. This process resulted in objects making statements of gender in different ways (Jahnke 2006), in which a common theme turned out to be a reflection on the design of seating. Postures and sitting behaviour are restricted by gender. As a man you are supposed to sit with your legs spread, stretching out while

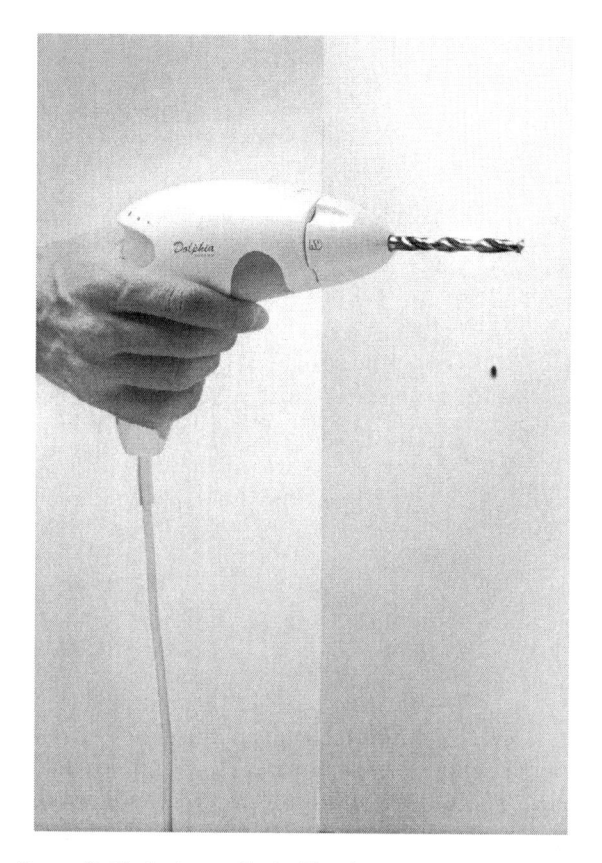

Figure 8.2 Power Drill; designer: Karin Ehrnberger; photo: Kent Johansson
Source: Courtesy of Karin Ehrnberger

a woman is expected to cross her legs. This can also be described in terms of space; a man takes space while a woman should make it. The male way of sitting and taking space is clearly visualized in the chair *Slothfully 2006* which aimed to comment on the male slacker. It invites you, despite your sex, to sit as a man, forcing your legs apart and to imitate masculine behaviour. *Stiletto*, the female counterpart, provides an insight in to how it feels to wear and balance on a pair of stiletto shoes, an attribute strongly linked to femininity and beauty. Wearing such shoes however also restrains the freedom of movement, which this chair illustrates, as it has only one leg and you have to balance this leg while seated. These two chairs complement one another but are also balanced by yet another chair, *Duel*, a chair with a freely turning backrest that forces you to cooperate, to give and take space, implying that we might have to work together to achieve equality. An exhibition ended the student project and it has been displayed repeatedly in different contexts since Summer 2006, allowing visitors to touch and test the designed objects, giving them opportunities to reflect and discuss gender with one another.

The design process itself turned out to be an important learning curve for the students as well, as they had to confront their own prejudices of gender and how to gestalt gender without falling into common pitfalls, and the reproduction of gender stereotypes turned out to be problematic. This insight is of great importance for every actor involved in the design and development of consumer objects who aims to create a more equal and gender conscious design. Raising the level of consciousness of how design is part of a process of gender construction amongst the public was another important aspect of this project.

This type of conceptual design is one way of instigating change. To most industrial designers a concept simply means to outline an idea, to create a product concept that can be produced and sold. However, conceptual design is a recent move by designers into art and politics, sometimes but not always resulting in a commercially viable product. Conceptual design tries to pull off effects of shock and sensation. With its capacity to experiment with different possibilities, conceptual design can proceed and create trends within a more flexible context. The transition to the commercial market is oiled. This way, design works as a means for preparing and initiating change (Buchanan and Margolin 1995).

The car industry uses product concepts and these concept cars are an important source of inspiration for realizing innovations in future car models. Concept cars(Figure 8.3) are not intended for production, and today's concept cars are more like conceptual design than product concepts. Every year a number of new concept cars are introduced to the car world, and in 2004 Volvo launched *Your Concept Car* (YCC), also known as the 'women's car' in Geneva. How it came about and how it was received will be revealed next.

A car by women for all

In Volvo's experiment (Figure 8.3) with a car aimed at female consumers all the described strategies for handling gender commercially were involved, and we will now explore the implications of this particular market segmentation approach. In 2001 Volvo Cars started the

Figure 8.3 Volvo YCC
Source: Courtesy of Volvo Cars

project to develop a concept car that would appeal to European women. All members of the project team were women: engineers, the project leader and designers, something that had never happened before. The YCC project started when Volvo invited Martha Barletta to attend a one-day seminar to talk about women and car design. The seminar had a great impact on some female engineers who thought: 'What would happen if women built a car?' The most memorable message of the seminar was, according to several of the YCC team members; 'if you meet the expectations of women, you will exceed the expectations of men.' They presented their ideas to the management who supported the proposal from the beginning. Volvo Corporation had been searching for ideas of this kind; the proposal suited the diversity and future-oriented spirit of the Volvo brand and also it was designed to target the growing female market for cars. The management had also been searching for ways of encouraging women employees to come forward within the organization and this was yet another reason to support the YCC project. The project was awarded a budget of around 270,000 euros – a fairly modest sum. The team claimed, in line with Barletta, that the project had nothing to do with equality; it was solely a commercial project. It was a successful way of motivating the project as a way of increasing the company's production that simultaneously undermined the critics who claimed that cars and equal representation have nothing to do with one another. It was, however, not entirely true, since the management did consider the YCC to be an equality project.

The car became Volvo's biggest marketing and media success ever. It was never produced and sold but the public relations value was enormous. Internationally the car was highly successful and many expressed interest in purchasing the car. The YCC was different from other cars with female purchasers in mind in that it was a prestige car positioned in the so-called premium segment and was not developed for a particular age cohort. 'Eve', the concept user, was a successful, urban European woman of no particular age. She was pictured as well-dressed and interested in cars, but not interested in maintenance. The car was given a sporty look and doors that open upwards, like wings. It was developed with a system called 'ergovision' in which each driver's bodily proportions are scanned to ensure that the driver has an optimal driving position based on personal measurements (see also Sparke 2005). Volvo's seats are normally adapted to suit tall individuals better than shorter ones and hence they tend to suit men better than women. From these details one could argue that the YCC challenges more stereotypes than the ones it confirms. However, the team encountered some resistance and also had difficulties in agreeing within the team on how to approach the idea of making a car for women. They discussed whether there were really any general differences between men and women drivers. Several of the team members were sceptical and it proved to be difficult to complete the initial ideas of a car for women. After an internal review the focus changed from being a car for women and it became Volvo YCC – a car designed by women, but intended for all. The marketing of the car had to be continually modified so as not to be judged as simplistically based on gender prejudice. Critical voices were heard not the least from inside the corporation where many Volvo employees described it as stereotypical. There were also discussions in the local newspaper that the car's design built on negative assumptions regarding women drivers. Design features that were often discussed were: the easy park system, the fact that the bonnet could not be opened and it was impossible to see the engine, also the windshield washer tank was filled by a capless ball valve positioned next to the capless fuel tank. Further, the seats had changeable covers, one of which had a pattern made up of embroidered flowers. The head-rests were designed with indentations to accommodate hair-styles like pony-tails and adjacent to the accelerator there was a fully adjustable heel rest to enable the car to cope with different shoe heel heights. The doors were designed so that it would be easy to climb in to the car without getting dirty. These details were highlighted by critics of the project as signs of a set of underlying presumptions concerning female drivers as vain or ignorant of mechanics. Some of the innovations later appeared in the design and marketing of other Volvo models, the C30 is visually similar and was marketed with the slogan 'think outside the box' to indicate a new approach to design. On the whole, however, the team expressed a great deal of disappointment in the fact that so few of the innovations were appreciated by the company's leadership. Many of them were, when we interviewed them, convinced that other car manufacturers such as VW would be willing to develop their ideas, unlike their own corporation.

On several occasions the project team experienced the difficulties of walking the tightrope between bringing invisible needs into the open and confirming existing stereotypes of men and women. The pony-tail adapted neck-rest was ridiculed at many presentations and later on in the process the team decided no longer to include it in the presentations. Alternatively, they embedded the verbal presentations in a narrative in which it was noted that many men had expressed an appreciation for the neck-rest which hence denied that it was based on anything feminine. The neck-rest is a good example of the ways in which, on the one hand, it is easy to dismiss a design because of the associations that are made between women, interior design, and superficial fashion statements. On the other hand, the neck-rest points to bodily variations through which an imagined normality is being created. The design of a vehicle reflects cultural discourses about the average driver, discourses that help reproduce the order. In that sense all ways of making them visible can be seen as good and help to disrupt and make trouble with conventional constructions of gender (Butler 1990).

Similar judgments were made of the heel support. Initially it was thought of as a practical and different design feature based on women's needs[2], but later it was modified and explained as building on an idea that the car should be able to be operated with high as well low heels. 'Women' as a category were played down further as the process went along. In the interviews the team repeatedly spoke of women as being more interested in functional details such as storage. The teams' associations with women and functionality were however contradictory and unstable. On one hand, they wanted to challenge such associations, but, on the other hand, they kept reasserting them. This indicates the arbitrariness of how gender relates to consumer products. It also suggests that these changes and complex contradictions can be used as examples of how consumer practises and the market make changes in the cultural constructions of gender possible. It points to the importance of seeing the world of commerce neither as passive nor as always negative. Even though changes may not always be for the better, they are not always for the worse and most importantly they signal the contingency of gender and markets (Laclau 1990 in Nixon 2004). Perceptions of the characteristics of masculine and feminine tastes are changing. The YCC intervenes in our discourses on men's and women's relationships with cars. It is powerful yet sensual and will not be defined as one or the other. It thereby takes action with the car industry's ideas of male and female when it comes to cars and produces a productive insecurity. The development of a car such as the YCC that is specifically adapted to female consumers can be seen as a way of bringing the hidden, male norms of the car industry out into the open and these cars confirm that women seek other values in cars compared to men.

Conclusion

The YCC project also points to the difficulties of working with gender issues in commercial design and the risks that are involved in this. It is easy to fall back on stereotypical notions that long since have been defined as feminine by the industry (Scharff 1991; Sparke 2003). In the interviews members of the design team noted that the car's enormous success had to do with the fact that it did not at all look the way the critics of the project had expected it to look. The flower patterned car seat covers are particularly interesting from this perspective. The process that produced them was everything but smooth. Most of the team did not like the flowers. They were seen as risk and as something that would place the team in the category of 'women designers'. The ambivalences regarding the flowers express insecurities around expressions of taste associated with femininity and a cultural awareness of the fact that different tastes have different values and definitions. Femininity is culturally the opposite of qualities that are expected from a prestige car such as power and performance. The team's ambivalent frames of mind also constitute feelings that are common amongst people who experience themselves as being a minority; that is, a fear of fulfilling those very same judgments of their own group that so often have been used to motivate their exclusion and to justify why they do not belong in that place (Puwar 2004). In retrospect, the team considers that the changeable seats were the most successful element of the car project, in the sense that it is where the company would have the greatest potential of taking market shares. Car seats and car interior have retained a feminine coding since the beginning of the 20th century (Scharff 1991), but the YCC project also highlights areas in which the gender codes of cars as well as attitudes to men and women have changed and are continuing to change.

We began by pointing out the benefits of taking a stand for a performative view of gender. Gender is made fixed through repetitions, in both actions and words. Here design, because of its materiality, becomes a possible source for introducing change and questioning assumptions. Design can show new possibilities, never before thought of, and can make concepts concrete and directly appealing to our senses.

Among designers professing to some theory of gender we found different positions. Some might support our own poststructuralist position, while others believe in a dichotomous model with men and women heavily polarized, a position we think is problematic as it does not facilitate the reception of new ideas, although they would think equality to be worth striving for – that is, the option of being separate, but equal.

The American marketing discourse literature has paid sustained attention to genders and markets and in our review of this literature we identified support for the argument that women's rights can prosper in a market environment. However, we disavow the limiting heteronormative understandings professed. Although highly critical of the theoretical gender sense underpinning this marketing literature, we still see some progress being made. Market feminism names things that never were named before and disrupts the sleepy routines of unstated male practices and discourses. A theory that is conservative at heart has let loose forces that are somewhat paradoxically producing more dynamic effects.

The section on design began by exploring the arguments of design theorists that suggest that often conservative routine practices can be broken, and this goes for gender and design as well. If only limited numbers of design projects exist in the field of gender they should be endorsed and supported. More inclusive forms of design are desirable; the standard of 'one design fits all' that is prevalent in the democratic idealistic design movements needs to be supplemented by a more diversified approach.

In this chapter we have developed a typology of strategies that we think are helpful in identifying the different approaches we observed in practice. First, we saw mixing at work, ideas driven to make society equal by making men and women more alike. Design serves to even out and ease the differences in clever ways, working to reduce the distance. Differences in some men's and women's bodies are explored in the second strategy, with gender theory intervening by exposing hidden conditions, like men being the unstated norm in the design of work clothes. Beyond the gendered body we found a third strategy that conceived the body in terms of size, rather than certain conceptions of men and women. Approaches four and five identified ways that can be deployed to enable designs, designers and people to cross the borders that are defined by gender. The substance of design associated with the activities of the opposite sex was made more acceptable by the articulation of highly visible symbols. Men got baby gear in camouflage, and women got power tools that were painted pink. The problem with this strategy is, however, that the overall tendency in society makes the objects of men more valued. The routine depreciation of the feminine side of the border must be fought. Finally, the sixth strategy involves breaking out of heteronormativity, the recent position of understanding gender through the truly inclusive position of queer theory, where the dichotomies are finally openly questioned. In queer design glamour is overflowing and any material gender expression will be available to anyone. But before reaching this overflow, queer design has left important impressions on design through playfulness and parody.

To conclude, we consider that the six strategies that we have tried to delimit and discuss make important contributions to the gender and design fields. To add brand value, companies should tap this exciting, high potential area and develop new markets.

Acknowledgement

Shortly after this chapter was completed our co-author Magnus Mörck passed away. He was a great inspiration and without his creative mind our project on gender, design and business organization would not have been as successful as it has been.

Notes

1. This point was also made by Marx when he suggested that a manufactured commodity must be used for it to have meaning. It is only through the process of consumption that commodities become real objects – see his Gruindrisse (1973: 90)
2. It must also be stressed that these processes were anything but smooth and there was a lot of controversy and disagreement in term of how to understand women and women's needs within the team.

References

Barletta, M. 2004. *Marketing to women*. Dearborn Trade Publishers, Dearborn.
Bloch, P.H. 1995. 'Seeking the ideal form: Product design and consumer response', *Journal of Marketing* 59:3, 16–29.
Brembeck, H., Ekström, K. and Mörck, M. (eds) 2007. *Little monsters: (De)coupling assemblages of consumption*. LIT Verlag, Berlin.
Buchanan, R. and Margolin, V. 1995. 'Introduction'. In Buchanan, R. and Margolin, V. (eds) *The ideal of design*. The MIT Press, Cambridge, MA.
Butler, J. 1990. *Gender trouble*. Routledge, London and New York, NY.
Butler, J. 1993. *Bodies that matter*. Routledge, London and New York, NY.
Catterall, M., McLaran, P. and Stevens, L. (eds) 2000. *Marketing and feminism: Current issues and research*. Routledge, New York, NY.
Cicolini, A. 2005. *The new English dandy*. Thames & Hudson, London.
Creusen, M. and Schoormans, J. 2005. 'The different roles of product appearance in consumer choice', *Journal of Product Innovation Management* 22 (January), 63–81.
De los Reyes, P. and Martinsson, L. 2005. *Olikhetens paradigm: Intersektionella perspektiv på ojämlikhetsskapande*. Studentlitteratur, Lund.
Du Gay, P. and Pryke, M. 2002. *Cultural economy: Cultural analysis and commercial life*. SAGE, London.
Dunne, A. and Raby, F. 2001. *Design noir*. Birkhäuser, Boston, MA.
Emilson, A. 2005. 'Design = Förändring'. *Konceptdesign*. Nationalmuseum, Göteborg.
Hannula, M. 2005. *Allt eller inget. Kritisk teori, samtidskonst och visuell kultur*. GU/ArtMonitor, Göteborg.
Hansson, L. 2006. *Universal design – a marketable or utopian concept?* Center for Consumer Science, Göteborg.
Jackson, P., Lowe, M., Miller, D. and Mort, F. (eds) 2000. *Commercial cultures – economics, practices, spaces*. Berg, Oxford.
Jahnke, M. (ed.) 2006. *NormgivningFormgivning*, Exhibition Catalogue, Center for Consumer Science, Göteborg.
Johnson, L. and Learned, A. 2004. *Don't Think Pink*, Amacon: New York.
Laclau, E. and Mouffe, C. 1985. *Hegemony and socialist strategy*. Verso, London.
Lee-Maffai, G. 2002. 'Men, motors, markets and women'. In Wollen, S. (ed.) *Autopia: cars and culture*. Reaktion Books, London, 363–70.
Loxley, J. 2007, *Performativity*, Routledge: London and New York.

Lurie, A. 1983. *The language of clothes*. Hamlyn Publishing Group, London.
Martinsson, Lena 2006. *Jakten på consensus*. Malmö:Liber.
Marx, K. 1973. *Grundrisse*, Penguin, Harmondsworth: Middlesex.
McCloskey, D. 2000. 'Postmodern market feminism: A conversation with Gayatri Chakravorty Spivak', *Rethinking Marxism* 12:4, 24–61.
Nixon, S. 1996. *Hard looks. Masculinities, spectatorship and contemporary consumption*. USL Press, London.
O'Connor, K. 2007. 'The body and the brand: How Lycra Shaped America'. In Blaszczyk, R. *(Ed)*. *Producing Fashion: Commerce, Culture and Consumers*', University of Pennsylvania Press: Philadelphia: 207–30.
Ostroff, E. 2001. 'Universal design: The new paradigm'. In Wolfgang P. and Ostroff, E. (eds) *Universal design handbook*. McGraw-Hill, New York, NY.
Partington, 1997. Partington, A. 1996. 'Perfume: Pleasure, packaging and postmodernity'. In: Pat Kirkham (ed.). The gendered object Manchester: Manchester University Press. 204–18.
Popcorn, F. 2000. *EVEolution: Eight truths of marketing to women*. Hyperion, New York, NY.
Puwar, N. 2004. *Space invaders: Race, gender and bodies out of place*. Berg, London and New York, NY.
Robach, C. 2005. 'Rapport från nuläget'. *Konceptdesign*. Nationalmuseum, Göteborg.
Scharff, V. 1991. *Taking the wheel: Women and the coming of the motor age*. Maxwell MacMillan International, New York, NY.
Scharff, V. 1992. Gender, electricity and automobility. In Wachs, M. and Crawford, M. (eds) *The car and the city: The automobile, the built environment and daily urban life*. University of Michigan Press, Ann Arbor, MI, 75–85.
Scott, L. 2005. *Fresh Lipstick: Redressing Fashion and Feminism*, Palgrave Macmillan, New York.
Sparke, P. 1987. *Design in context*. Bloomsbury, London.
Sparke, P. 1995. *As long as it is pink. The sexual politics of taste*. Rivers Oram Press, London.
Sparke, P. 2003. 'The "ideal" and the "real" interior in Elsie de Wolfe's the house in good taste of 1913'. *Journal of Design History*, 16(1), 63–76.
Sparke, P. 2005. 'Volvo's YCC Your Concept Car'. *Torsten and Wanja Söderbergs pris 2005*. Röhsska Museet, Göteborg.
Spivak, G.C. 2000. 'Other things are never equal: A speech', *Rethinking Marxism* 12:4, 24–61.
Thackara, J. 1988. *Design after modernity – beyond the object*. Thames & Hudson, UK.
Welch, P. 1995. *Strategies for teaching universal design*. Adaptive Environments Center, Boston, MA.
Woodham, J.M. 1997. *Twentieth-century design*. 141–164. Oxford University Press, Oxford.

9
Riding the Waves of Design: Industrial Design and Competitive Products in the Norwegian Marine, Maritime and Offshore Sectors

Grete Rusten

Introduction

Industrial design as a tool to enhance competitiveness has received increased attention in many economies. Many Western countries have begun to develop policies that are intended to enhance the design-intensity of products that are produced by indigenous firms. Industrial design is also considered by the Norwegian government as a strategically important tool that brings added value to new and existing products in ways that enhance the ability to support a higher price. The latest design policy initiated by the central government calls on producers to increase the awareness, knowledge and use of design to develop better and smarter product and process solutions. Priority has in this policy document been given to the maritime, tourist, food and forest industries (Nærings- og handelsdepartementet 2008). The programme that followed up on this declaration began with a budget of 10 million Norwegian kroner (NOK) in 2009, and includes three tasks: a national business survey to investigate the use and competitive effects of design, economic support for new design projects and a Design Effect Award organized by the Norwegian Design Council (http://www.norskdesign.no).

This chapter explores the relationship between industrial design and competitiveness through a detailed analysis of the Norwegian maritime industry. I have especially focused my study on cases that involve building ships using yards in Norway. The building of ships in Norway includes mainly specialized vessels such as naval, research and commercial fishing vessels, as well as offshore vessels and passenger ferries with a high technology content. Other important categories within the maritime industry are ship-equipment, (production, trade and installation), design, engineering and other supporting services related to this sector. The maritime industry is a suitable case study industry for two reasons. First, the maritime industry remains an important sector within the Norwegian economy and, second, this is perhaps the first analysis of design in this sector. Research into design-intensive products has tended to concentrate on the analysis of the fashion industries, for example, furniture or the automotive sector, and has largely ignored sectors like the maritime industry. The aim of this chapter is to identify and explore the business models, organizations, products and geographies related to industry-oriented design projects that are aligned with the maritime sector.

A key question that will be addressed is how can design be of strategic importance to a business? There are several contributing factors. A practical and appealing design can increase sales, as satisfied customers recommend the product to others. Design can also add value to products and can transform a standard design with a low value into a fashionable high added-value good. New markets and more efficient production systems can be additional benefits of investment in design, thus enhancing profit levels. Product differentiation in the marketplace may also be achieved by the development of products that can be customized to suit the needs of individual purchasers.

Design turns ideas into practical and attractive solutions (Best 2006). Factors such as identity, aesthetics, history, experience, functionality and environmentally friendly processes, together with product solutions related to content, use or recycling are important. Such issues will, through a focus on design, have a place in the product development strategies of many firms which supply goods and services to the industry, professional clients or private consumers. A functional and attractive design may also be a way in which brand names can be established that support or have been created from the development of recognizable products associated with the company. The success of a product may further depend upon finding powerful demonstration arenas or marketing spaces, such as the catwalk, the reception area of an office building, an airport, a hospital, a ship, a production floor, media (fashion magazines, product placement in films and television, etc.)

and other places where people meet, work and exchange information. Companies that actively integrate design as part of their business strategy now often choose to follow the product through all its stages, from project development through the complete value chain until the moment of final consumption. Often industrial design was something that was bolted onto the research and development (R&D) process, but in many companies it is now completely integrated throughout the product development process.

How can one actually describe industrial design? In a technologically oriented production context, industrial design can be considered as part of the wider process of knowledge input and material development that may also include R&D and engineering. In these processes, one can distinguish between industrial designers and engineers by their formal training background and professional associations (Rusten and Bryson 2007). In addition, there are often important differences in the legal contracts that surround design as compared with engineering especially when projects are arranged through external engagements. Designers frequently have a looser contract arrangement than engineers and researchers. Many designers have contracts that are based on royalty payments and the number of products that are sold rather than being fixed-price agreements that are independent of sales (Kristensen and Lojacono 2002). A further difference distinguishing design from engineering is the emphasis placed on the development of designs that meet the needs and preferences of targeted customers (Farstad 2003). A third difference between designers and engineers is the approach and methods they use in their work. Designers are inclined to think of a setting related to the environment of the user. They are, in other words, more oriented towards the customers than to either the market or technically based innovations. The object of being in contact with the market is to both promote the product as well as to receive feedback on how well it works and is being integrated into the lifestyles and activities of consumers. Such information may be useful in making modifications to existing products as well as in the development and design of new products.

The empirical evidence in this study is based on a larger research project on design economies that included the analysis of secondary statistical data and 25 in-depth interviews (16 with client firms and nine with external designers). Additional data from key business informants, earlier studies, websites and magazines have been part of this study. The interviews were undertaken during 2005–6 and cover a range of industrial sectors in Western Norway of which six were mainly involved in the production of products targeted at firms and industrial markets.

The method chosen has been a useful way to increase understanding and analysis of the strategies and complexities that surround the relationship between industrial design and competitiveness. It has also highlighted design competencies that would be difficult to identify using a more quantitative approach. Five design cases based on the full collection of sources are presented in this chapter. A multi-methodology was adopted that involved triangulation between secondary sources, empirical evidence, case studies and web pages. The emphasis is on exploring the role industrial design plays in the development and commercialization of intermediate products – products developed and designed for incorporation into other production processes.

The geography of design expertise: A theoretical perspective

Much of the focus on design referred to in economic geography is linked to literature on creative communities in larger urban centres (e.g. see Florida 2002; Scott 2000; Pratt 2002). This literature analyzes creative communities clustered in cities such as Los Angeles, Paris, New York and London that are working with fashion, music, dance, film, art and design. Some of the exhibitions, theatres and shops found in these locations are the result of real estate and city culture projects that have replaced traditional manufacturing activities. A renovated flour mill in Newcastle, a brewery in Los Angeles, a shipyard in Oslo and a canning factory in Bergen are several examples of situations where cultural or consumer activities have been placed in buildings that were originally created for manufacturing activities. The locations of the designers in these urban places means that they often are geographically separated from their client firms since many producers are located elsewhere in the country or abroad. A geographical distance between design service providers and customers is also observed in several of the maritime design projects that have been identified in this study.

In the literature on the creative class, creativity is something that is considered to be located in large urban cities and this raises the question of whether the relationship between creativity and urban spaces is relevant for a country like Norway (Montgomery 2005). The answer is perhaps both 'yes' and 'no'. Yes, because some of the designers in Norway have identified themselves with these urban lifestyles and have moved to London, Stockholm, Paris or New York. No, because other designers have sought an alternative lifestyle to busy metropolitan life and established a home in smaller cities or towns in Norway. Eighty per cent of Norwegian design firms are located in the five largest city regions, but these centres are hardly comparable with the larger cities that the creative class literature refers to. The remaining 20 per cent are located in small towns and rural areas (Rusten 2008). Additionally, what about the majority of designers who are employed in-house by manufacturing or other businesses and are geographically distributed over the country? There is perhaps an over-emphasis in Florida's work and in the literature on the creation class in large urban areas and a neglect of design activities that occur in other places and spaces.

A further critical argument that limits the relevance of the creative class literature to this study is associated with the location decisions made by designers. The location of some design clusters in Norway is not determined by social or cultural factors. A relatively decentralized educational system and industrial concentrations, for example, related to the maritime, furniture or oil industries, means that Norway has a relatively dispersed pool of talented individuals, but also a requirement for design expertise outside the main urban areas. Some of the industrial clusters are also the result of regional, private or public real estate and cultural development initiatives. A closer inspection at an individual level also reveals that some of the location decisions are a result of marriage, partnership, a search for outdoor recreation or other personal or behavioural motivations (Rusten 2008). Good broadband coverage and well-developed airline connections also make it possible to live outside the cities and still effectively reach national or overseas customers when necessary (Rusten and Bryson 2007).

How the use of design affects the geography of production systems

The discussion so far has been about geographical context and scale. A further comment is connected to geographical range. Florida (2002), for example, through a focus on individuals, refers to the geography of talent and social organization connected to specific urban clusters. It misses the point that many categories of expertise are attached to production systems that are considerably geographically dispersed. An example of a work that covers the spatial division of labour is Dicken (2007), who considers models of transnational production units with particular reference to electronics, cars, textiles and clothing. Yet other studies refer to the geography of knowledge through R&D (e.g. Fuchs 2005; Coe et al. 2007) or spatial division of expertise (e.g. Bryson and Rusten 2006). These are amongst the literatures that analyze the complexities of production units and processes that are linked across national borders.

Industrial design related to the spatial division of expertise does also reveal that competence can be related to geographical effects in different ways; effects to be considered are as follows.

1. Which parties and locations are involved in the design process?
2. Where are the designs 'consumed' through use of the finished product?
3. How do certain designs become important elements for the survival of production units even in high-cost locations?
4. What design styles, such as module detail, enables a vertically linked production connection between locations in different countries?
5. How may certain products be associated with and recognized by their geographical origin?
6. How is it that some products are designed in one country and produced in others?

These components will be further explored when considering the presentation of industrial design within the maritime sector. Designers require a detailed knowledge of the company, including its image, business concept and market position. There is also a need for expertise in the product field, including price levels, quality, style, technological development, regulations and other market conditions (Monö 1997).

This chapter embraces some of the complexities related to industrial design and its role in the production system by referring to specific products and business sectors. This therefore presents a different focus from the classical creative industries literature where the categories of design either refer to designers in general terms or deal with fashion and consumer goods. Hitherto, the ways in which designers are organized and their work with industry have not been an important focus in much of the creative class literature.

The design process is all about understanding and responding to present and future needs and preferences. To achieve success there must be a close social proximity between the designer, industry and the consumer or user. Geographical nearness can be considered convenient rather than absolutely necessary. Interpersonal qualities such as the ability to listen, observe, generate a vision and communicate are important skills for designers in their work with firms. Being in close contact with customers and having an understanding of their interests and needs determines the success of these projects.

Systematic investigation into the situations of different designers engaged in projects within the maritime sector suggests that there are a variety of organizational models that encapsulate the design process. External designers may be organized as individual consultants, or engaged by larger companies specializing in design or by selling services in combination with engineering and other expertise, possibly integrated with production. External designers may also be either dedicated to specific sectors or may be able to cover a broader customer and product portfolio.

As stated previously, the success of projects is determined by maintaining close contact with customers, and understanding their interests and needs during the process. Solutions to problems are found more efficiently and they are more creative and innovative when the parties meet face-to-face (Sweeney 1996). The fieldwork methods of industrial designers are also very much about observation, dialogue with immediate feedback and testing. Informal working conditions and close contact between designers and industry are important to be able to create designed products that are both aesthetic and functional. An industrial designer sometimes will need to undertake a field visit to the actual users of the product. It is therefore all about the geography of projects rather than where design service providers and clients are located. Valuable information for the design process is picked up through observation and conversation. A field visit example that illustrates this is when designer Einar Hareide follows an ambulance boat to observe how the crews work together and the space and equipment they use (Rusten 2008). Close customer relations may lead to the generation of new sales as well as enhancing product performance and in securing customer referrals to new potential customers.

The focus has been thus far upon the designers' location and where they work. Geography can however, also be linked to products. The maritime industry's products, for example, are designed to be able to withstand rough weather conditions, and are known in the market to have these qualities. Some Norwegian ship and marine equipment manufacturers have a leading international market position for the production of specialized vessels often operating under heavy weather conditions and in difficult waters. Such conditions create a competitive element in world markets (Rusten 2008).

The geography of design: An empirical approach

There are methodological challenges associated with the identification of industrial design. Design is often identified by sector, but design may also be an invisible or hidden activity and may be difficult to identify. This methodological challenge became evident when analysing industrial design using Statistics Norway's Central Register of Establishments and Enterprises (CRE). The CRE is a comprehensive register based on direct contact with the units and use of the National Insurance Administration Annual Check Database. Design firms in this database have been identified using the NACE classification (Nomenclature générale des Activités economiques dans les Communautés Européenes code 74.872, Statistics Norway 1994), but this covers only those designers that either have their own firm or work for businesses in which the main commercial activity is classified as design. Companies that also define their major activities as the production or trade of goods are not registered for design, since this definition only serves for those companies selling design as a service. The most recent data are from 1 January 2006 and identify 2756 industrial design companies, most of which are very small. As many as 2444 companies (89 per cent) are registered as sole practitioners and of the remaining 312, only 50 are shown to have more than 4 employees and none have more than 50 (Statistics Norway 2008, unpublished data).

The CRE dataset shows the number of industrial design firms and their size structure measured by number of employees regardless of their profession or role. These data show the number of design firms, but not the number of designers. An alternative measure that will also include those designers working in sectors other than the design industry is to use data that indicates the profession or occupation of the employee and that are sorted by workplace sector. The Statistics Norway count of professions for 2006 provides the figures for designers that are hired by production companies. It also includes those cases where designers themselves establish a design business by either setting up their own production line (this is classified as manufacture) or by purchasing such a capacity and then selling the designed products themselves (this is classified as trade). By including sole practitioners in the database, a list of 7772 designers is derived, of whom 35 per cent are sole practitioners, 16 per cent are designers working in firms registered as design businesses, and 49 per cent are designers working in firms in other sectors (Statistics Norway 2008, unpublished data). Some comments related to the industrial structure of the design industry in Norway are presented in the following sections.

How being small represents a problem when operating in a knowledge-intensive market

The dominance of small design companies has a number of interesting and important implications. Many small design firms face a considerable challenge related to how the design function is considered by potential client firms. They also have great difficulties in balancing limited resources and capacity with the result being that limited time is available that can be devoted to the strategic development of their business and to the cultivation of new or even old clients. There are several reasons for this.

1. Small design firms have limited personnel resources; this means that they often lack capacity and in consequence find it difficult to follow up all potential client enquiries.
2. A further challenge concerns the content of the service that is offered. Designers must obviously know the industry they are working with to be able to develop a good product or service that will be accepted by specific business clients and will be attractive to the market. However, with few staff it can be difficult always to be fully up-to-date.
3. The designer needs to acquire current information about safety and quality regulations and environmental issues, both in general and related to specific products. A small company will often face difficulties when it comes to its capacity to handle all the relevant information. Nor will it be easy to assemble staff where each member specializes in undertaking certain tasks.
4. A small design firm may also find it difficult to be able to follow up and adjust designs related to a new generation of products that have recently reached the market.
5. Many small firms also find themselves much occupied with routine tasks and may have difficulty in finding time for acquisitions. The tasks mentioned are very time consuming and mean that many small design firms are engaged in very few projects at any one time. That can be a risk in a market situation where the competition is fierce.

Being in a team with others is not just a way of solving capacity problems; it can often also be more inspiring than working alone. One of the maritime sector design teams that we interviewed referred to the ways in which creativity was a mix between developing something individually, and being able to share ideas, discuss them in detail and improve them with other team members. He referred to projects in which one of them had designed an important facet for a project vessel and the designer was able to call upon other members of the project team to have a look at the concept, give an immediate response and perhaps make suggestions for improvement.

The discussion so far has dealt with the designer as a service provider and his or her working conditions, but the issue about the size of a firm also has to do with how the buyer sees these small organizations. The petroleum and maritime sectors are examples of industries with many high-cost investments in which the choice of suppliers is based on very stringent criteria about size and capacity as a way to minimize delivery difficulties. They also have comprehensive requirements with regard to the suppliers, which include evaluation criteria connected to customer references, certificates and product documentation standards. These companies will simply not engage suppliers when the delivery of a product or service is totally dependent on one person. It may also be about what is expected to be included in the delivery. In some cases, for example, the customer requires a project management team as part of the project. This means that the designers will have to put together all deliveries and follow the design all the way through the production system. Having all these tasks delivered through one contract will again favour design teams with a larger and more varied set of competencies. Our experience from research into industrial design and projects has shown that not all designers have the necessary project management training (Rusten and Bryson 2007). On these grounds, therefore, some of the smaller design firms in the market will be excluded from some potential design projects.

Degree of specialization

To be identified as a potential supplier means that a firm has to be well known and recognized in the marketplace. It is also a question about whether it is best to be a specialist within an industry or to be more of a generalist and able to cover a broader field. Some designers who have projects within the maritime sector have had an earlier career in the maritime sector. They also take advantage of other categories of expertise such as marketing, engineering or other specializations that they may have amongst their own staff or through suppliers and other collaborating parties. In some cases, regional, national and international research collaboration forms part of the programme package.

Of course, there are also the important considerations of competence, communication and a reputation for the delivery of projects. Being visible in a sector and being associated with a specific project provides a designer with a reputation that can be assessed by potential clients. Being associated with a specific sector may be one way of establishing a name or reputation in a specific industrial sector with the result being that the design company will be included in invitations to tender for specific projects. Too much dependency on a particular sector may, on the other hand, enhance vulnerability and dependency, and may also mean that the design company is exposed to any recessionary downturns that their core client based is exposed to. Generalization is associated with a set of advantages related to the development of a reputation and client contact network, but it also has important disadvantages. Generalists find it difficult to develop a reputation in a particular industry and may find it difficult to grow rapidly without developing a sub-specialization within the framework of a design firm that offers services to all companies and covers all sectors.

General design firms have advantages related to economies of scale and also learning that can be transferred from project-to-project. These opportunities to transfer learning between sectors and clients can concern not only design details, but also other elements such as the way projects are organized. For firms that have a long experience in working with the petroleum sector and have high skills and standards in the production of quality documentation and management, such refinements have proved to be very useful when working for other sectors. The designers sometimes also see product possibilities in new industrial markets that the firms themselves have not considered. Surveillance cameras used for underwater petroleum operations can, for instance, also be used for the inspection of aquaculture installations, sewers and cables, or for maintenance purposes on ship hulls (Rusten and Stensheim 2007). Designers can also identify products that could have other applications and in that manner work to encourage the development of new business partnerships.

The organization and geography of projects

It is important to actively involve client firms in the various stages of the design process. A design firm that consists of several individuals with different specialties working together in the same location where they can easily meet customers is an undoubted advantage. It is actually all about mutual understanding. Design teams and designers who have had careers in or project experience with shipping seem to have an advantage as they know people and the culture in this sector. On the other hand, Designers who have acquired most of their experience working in other sectors may possibly bring new knowledge and understanding into a project. In fact, having a project team that includes both these categories of experience seems ideal.

The design process may also involve experts with an artistic background. Sculptors may be required to model a prototype in clay.

Decorators may be engaged to work on the interior details of a vessel. Design projects rely on multiple information sources including personal experience, observation, communication and written material. The provision and reception of information is not straightforward. Both the industrial design service supplier and the client buyer of this service provides valuable information to the process.

The issue of how design knowledge is organized tends to depend on whether a company relies on internal or external resources. This polarized categorization can easily be regarded as being too much of a simplification. The nature of knowledge exchange in design and innovative projects often combines internal and external design resources. In cases where the design project is managed by client firms, they will often also supply teams with other categories of expertise, such as in engineering, management or marketing.

The geography of the maritime sector

Norway has the fifth-largest maritime fleet in the world as measured in terms of ships and registered tonnage, and this is represented by most of the large shipping categories including tankers, bulk transporters, chemical, container, refrigerator, liquefied natural gas (LNG), offshore and car carriers. The maritime sector in Norway, including the ownership of vessels, rigs, ship and rig building yards, equipment producers and maritime services, is the second-largest industry in Norway (http://www.maritimt-forum.no) after the petroleum sector. The gross value in 2008 was measured at 100 billion NOK with 90,000 employees (http://www.ncemaritime.no). Norwegian-owned shipyards have responded to global competition in different ways. First, they have concentrated on the construction of ships that require specialized technical details, equipment and functions. These reflect markets in which there is reduced price-based competition. A niche orientation also means that the firms have the ability to recognize nuances in needs and individual preferences in differing geographical markets. Secondly, they have outsourced the building of hulls to low-cost locations like Romania, the Baltic countries and Poland while the management, design and main part of the fit out is undertaken in Norway.

The Norwegian shipyards have developed a spatial division of labour that is, in part, related to the development of flexible production solutions that efficiently tie together the different parts of the operation. The maritime sector can be divided into two main parts, shipping and shipbuilding, and is mainly concentrated along the coast of Western

Norway, parts of the south and Oslo, but also with some activity further north. The major shipping town measured by tonnage is Bergen. About one quarter of workers in all aspects of the production system are located in Møre og Romsdal county, which has been given the status of the Norwegian Centre of Expertise for the maritime industry by the Norwegian Government (http://www.ncemaritime.no). The largest shipyards are located in Møre og Romsdal.

The maritime sector that is principally oriented towards the petroleum industry is mainly located in Sunnhordland, south of Bergen, and in the Stavanger area; other smaller towns have firms oriented towards the fishing fleet. Some places have yards and services oriented towards repair whilst others build new vessels. A new promising specialist niche that has been developing is oriented towards the offshore market and the development of submarine technology. The regions from Rogaland (Stavanger) to Møre og Romsdal in western Norway together make for a relatively complete maritime cluster, with many categories of product supply and services both to the civilian and naval sectors. The linkages are partly to regional, but also to national and international markets (Reve and Jakobsen 2001). The majority of headquarters and competency-intensive production activities seem to be relatively secure in their existing locations both in larger urban areas and smaller towns along the Norwegian coast.

Competences, organizational models and market diversity related to maritime design projects

The Norwegian Design Council database (Feb. 2009) lists 43 designers who identify the maritime sector as their primary market, but many of the naval engineering and design companies are absent from this list. The incomplete listing is partly to do with the fact that the database has recently been developed, and requires the firms to actively register themselves. Some of the ship design and engineering companies have probably not registered themselves with this database because they are already listed elsewhere. Examples are registers based on membership of different industrial federations, for example, Norsk Industri, Maritimt Forum and the TEKNA Society for Maritime Technology, as well as market- and trade-oriented databases dedicated to the maritime and offshore sector. The fact that many designers are hired as in-house staff in the maritime sector makes a further complication in measuring the full scale of this category of expertise. Then, of course, there are all those designers who are working with other categories of business. The number of designers who could be counted as targeting the maritime sector would further depend on whether one included only those designing vessels or also all those dealing with equipment and interior products. Finally, one has to decide whether the basis for the count is the existing project portfolio or whether it also includes the market ambition of the design company.

Different categories of design firms and teams

The following nine examples were identified through our research into industrial design in the maritime sector; they illustrate the large variety of companies that sell industrial design as services or products to the maritime sector. Some structural characteristics are shown in Figure 9.1. The nine examples represent different organizational models for design projects:

1. Hydrolift: a company producing high-speed boats that are based on the designs of the industrial designer–owner Bård Eker.
2. Ulstein Group: a maritime industry company with an in-house design team. This company covers the shipbuilding, equipment and shipping services sectors.
3. LMG Marine: a specialized naval architecture and engineering company, and provider of technical services.
4. Vik-Sandvik of Norway (from mid 2009 ownership change-now Wärtsilä Ship Design) Group: a specialized naval architecture, design and engineering company concerned with the design of on- and off-shore equipment.
5. Design and communication service companies that specialize in the maritime and offshore sectors (e.g. Maritime Colours: a communication service company that specializes in the maritime and offshore sectors).
6. Hareide Designmill: a design service company working with various categories of products, services and clients.
7. CMR Prototech: a company that is part of the University of Bergen, specializing in research, engineering and product design and operating across industrial sectors.
8. Maritime Møbler: a specialized furniture producer.
9. Orbit Aquacam: an underwater camera manufacturing company producing designs and products for various industrial sectors including the maritime industry.

The design of products for the maritime industry is very often an interactive process that benefits from the competence and vision of different categories of expertise, including those of the customer. All the companies on the list have in-house designers. The teams sometimes also rely on expertise hired on a project-by-project basis.

These cases illustrate the ways in which designers and design firms have chosen different market strategies. Some specialize in just one product category or industrial sector whereas others operate with many different products, clients and industrial sectors. Five of the nine companies specialize in selling design services. Of these, companies (3), (4) and (5) specialize in the maritime sector, whereas companies (6) (Hareide Designmill) and (7) (CMR Prototech) are more generalist providers with respect to types of products they cover. The designers at Hareide Designmill work on a wide range of designs including cars, footwear, boats, wood-burning stoves, snowmobiles and hunting knives. Amongst the products that CMR Prototech has designed are a telephone system for tunnels, and a flowerpot for use in a space shuttle expedition. Firms (8) and (9) represent two categories of suppliers. The first has a market orientation towards the maritime sector whereas the second covers a wider product market range. Both have used external designers.

All the companies except for Hydrolift are located in the maritime cluster in Western Norway, both in the city area of Bergen as well as in smaller urban and rural areas. The following sections explore these examples of design firms in more detail.

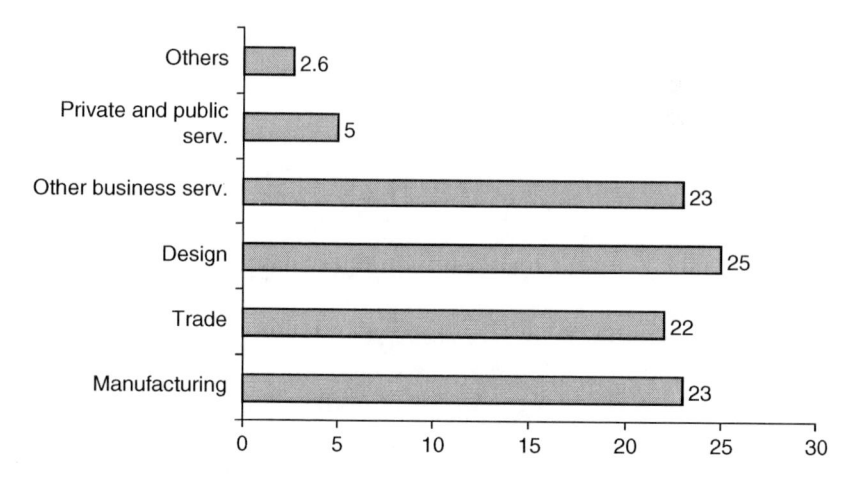

Sector	Value
Others	2.6
Private and public serv.	5
Other business serv.	23
Design	25
Trade	22
Manufacturing	23

Figure 9.1 Distribution of designers by sector (Statistics Norway – unpublished data N=5016.)

Note: Figures do not include sole practitioners

Product design examples connected to the maritime industry

The chapter now turns to a closer inspection of some case studies to illustrate project examples and the different ways in which industrial design is incorporated into products and production systems. The first examples refer to designs that in addition to providing functional and smart products, include an example in which design was instrumental in rationalizing a production process. A good example is the development of a modular system, which is a strategy that combines customization and mass production.

High-tech processes use robots and product elements designed as modules, which enables a firm to produce product variations but also to ensure production costs remain relatively low. A larger sales volume can more easily cover the design costs one has incurred in the development of the modular system before it was offered to the market. A further efficiency gain may be in the functional detail that will simplify the installation process. It may also become much easier to replace or add elements when needed. One product example from the maritime sector is a bridge control station system based on console modules. This product, designed by Vik-Sandvik of Norway (from mid 2009 ownership change-now Wärtsilä Ship Design) Industrial Design, offers a flexible bridge with customized details with sections that can be easily replaced, added-to and upgraded when required (http://www.Vik-Sandvik.com). Another example is MaritimeMøbler, a leading producer of maritime furniture, which launched a module-based furniture collection consisting of standard parts that could be fitted together in different combinations. The company hired expertise from Vik-Sandvik to develop this design. This new way of production increased the firm's efficiency by 50 per cent and reduced the costs for the customer (Norwegian Design Council 2007).

Another industrial design example from the maritime sector is the AutoChief C® 20 control system, designed by Hareide Designmill for Kongsberg Maritime and produced in Norway. This steering system, shown in Figure 9.2, is designed for medium- and low-speed diesel engines that are used for larger ships where the functionality is simplified and optimized. The modular design used in this product makes it easy to meet individual customer requirements. Robust materials and few moving parts also reduce maintenance costs (http://www.km.kongsberg.com). This project received the Norwegian Design Councils Award for Design Excellence in 2003 (http://www.norskdesign.no).

Design can also be about creating emotional values. Designing a smart-looking instrument panel for a vessel is a way of creating a modern and attractive workplace that will be important for future

Figure 9.2 The Autochief ® C20 propulsion control system for maritime vessels is designed by Hareide Designmill and produced by Kongsberg Maritime; the design received the Award of Design Excellence by the Norwegian Design Council in 2003

recruitment. Being staffed by employees that are satisfied with their working environment also has a positive impact on productivity and safety.

Some ships are also constructed following a modular-based project management approach. For instance, Ulstein develops specialized vessels. Examples are offshore, multipurpose, chemical carriers, container cargo and RoRo vessels. The project was initially one that included Ulstein's in-house design unit and the Oslo industrial designer Emil Abry. The project was supported financially by the Research Council of Norway and other external partners (Norwegian Design Council 2005). The combination of a flexible/standardized design with serial production provides economies of scale related to both the design process as well as construction.

The presentation of examples so far has focused on smart design solutions that not only enhance the efficiency of the production process but also improve the effectiveness of the product for the final consumer. The following examples will deal with functional and distinct product details.

The ULSTEIN X-BOW® containership shown in Figure 9.3 has a unique appearance. Its characteristic bow shape allows a faster service speed

Figure 9.3 The ULSTEIN-X-BOW ® Containership. Ulstein is introducing a broad a broad range of innovative and cost effective short sea vessel designs

and more reliable operation, as the sailing schedule can be kept even under rough weather conditions. Additional advantages include reduced slamming and green-water spillage on to the cargo and more comfort for the crew on board. The ULSTEIN X-BOW® shape also reduces icing. Lower fuel use and reduced emissions are important environmentally (Ulstein 2008). The testing and teamwork between designers, ship engineers, maritime personnel and shipyard workers have resulted in this unique bow concept. Four vessels are already in use, and 29 others are under construction (Nærings- og handelsdepartementet 2008).

Environmental effects can also be linked to five LNG (liquefied natural gas)-fuelled ferries (Figure 9.4). The project was initiated to increase the transportation capacity on some busy ferry stretches in western Norway (Karlsen et al. 2005). Replacing diesel fuel with LNG and consequently lowering the emission of CO^2 by approximately 25 per cent and NO by 85 per cent has also been important. The NO reduction was equivalent to the total emission of this gas from all the cars in the capital city Oslo. Although this project was launched mainly for transport and environmental reasons, it additionally was one of several initiatives to use more natural gas. The innovative use of LNG-fuelled engines represents an important demonstration of efficiency that may lead to the export of vessels, engines, equipment and services by the Norwegian maritime sector. Transport along the North Sea coast and other short

Figure 9.4 LMG Marin of Norway and this LNG fuelled passenger ferry is based on design and technical configuration by LMG Marin AS of Norway and operated by FIORD1; replacing conventional diesel fuelled ferries has reduced emissions and increased the ferry capacity on some fjord stretches

Photo: Harald M. Valderhaug

sea stretches as well as ship transport on rivers in Europe represents an interesting market for LNG-fuelled vessels (Karlsen et al. 2005).

The new 21-knot ferries with a capacity of 250 vehicles and 600 passengers are faster and larger than the previous generation of vessels operating on this route between Halhjem and Sandvikvåg on main highway E39 between the cities of Bergen and Stavanger. The design and technical configuration were completed by LMG Marine, Bergen. Aker Yard won the tender bid and built the vessels using yards in Romania to build the hull, but the internal fit and equipment were undertaken in Søviknes and Brattvåg, in western Norway (information from LMG Marine). The ferries have been designed in accordance with the requirements of the customer, national regulations and the environments in which they will be used. The change of fuel, with its whole range of technological solutions, storage capacity, security and functional details, has given the vessels a new style. Details such as larger saloons, Internet access and more windows in the passenger area also enhance passenger comfort (Skipsrevyen 2007). Five ferries fuelled with LNG are planned to increase the capacity on some of the busiest stretches between Bodø and the Lofoten archipelago; they will come into operation starting in 2012 (Avisa Nordland, 17.02.2009).

Yet other designers specialize in developing designs that use advanced materials. One example is the firm Umoe Mandal, which specializes in the design and production of advanced high-speed vessels, mainly for naval purposes. These vessels are based on highly specialized, low-weight composite.

The analysis has so far focused on industrial designs by referring to some physical products. Design services oriented towards the maritime sector can also be linked to marketing and communication. An example is Maritime Colours, founded in Bergen in 2002, which now has 31 employees (2009). Maritime Colours' major market is in the maritime and offshore sectors in Norway and in international markets; it provides a full range of services from conventional marketing materials to interactive web-based communications solutions. The 3-D technologies developed for computer games are used to visualize and simulate complex maritime products in a more accessible manner. It is also possible to navigate around an image to show different angles and technical specifications in presentations that can be used for conventional marketing publications or as web-based communications for marketing, information or training purposes (Bergen Chamber 2005). The use of 3-D modelling and animation is also considered to be a useful tool when bringing together different areas of competence to study models during the design process.

Design education

In a global economy firms located in developed market economies with high employment costs (salaries and social benefits) find it difficult to compete with firms located in countries with low labour costs. One solution is for firms to shift away from price-based competition towards the development of products with high-added value. This is the strategy that has been followed by the Norwegian maritime industry. This business strategy is only possible if firms are able to balance and coordinate aesthetics with the commercially oriented aspects of a business.

It seems crucial, however, that the designers have to understand the industry and its specific values and needs. Managers must understand and appreciate the value of design in all stages of a production process;

from the initial planning all the way through manufacture to the finished product that is introduced to the market. One even has to accept that some projects will not succeed. The production of high added value products requires a balancing of two types of knowledge: design versus commercial knowledge. There is a danger that a designer will create a product that has limited profitability, or that a company may reduce the design features of a product to enhance profitability and at the same time destroy or undermine the commercial advantage that might come from the designer elements that were incorporated into the design. Design education is extremely important in this context, especially an education that blends training in design processes and aesthetics combined with training in business and enterprise. The purpose of such combined training is not that the designer should end up with a full business degree, but it would go some way towards ensuring that designers appreciate commercial logic which might ensure that more designed products reach the market.

What seems important in relation to the Norwegian economy, which has so many small and medium sized enterprises, is that industrial designers understand how to organize a design project. Many smaller client firms do not have the capacity to undertake this task. Organizing a design project through all its stages will simply be one element of a contract. Another element that may be crucial to the industry is how to understand the nuances of different geographical markets. A product with a particular shape and functionality may sell well in Norway, but not necessarily in other places in Europe or more distant markets. Understanding cultures and how they affect styles, designs and brands thus also seems to be important.

The designers must know the industry they are working with in order to be able to develop a good product or service that will be accepted by specific business clients and which will then sell in the market. A successful industrial designer needs to have both an artistic and business-driven mindset that fully understands the culture and work practices of the maritime sector to be able to work with these customers.

This requires training that includes the business perspective with a focus on understanding competitiveness, market orientation and internationalization. Courses that analyze business strategies and the global economy will have to be included as part of the formal training of designers. A design service from idea to product includes the identification and definition of roles and responsibilities, and may include many parties both as individuals and subcontractors. Elements that have to be put in place and coordinated throughout the project include materials, skills, equipment, location, processes and standards.

The designers we have interviewed often noted that their formal training did not provide them with the knowledge required to run projects within businesses. An arena where experienced senior designers work together with newcomers, or alternatively where designers collaborate with management consultants or business experts are two ways of meeting the challenge of following a project safely all the way through the process.

Conclusion

There is a tendency for academics to focus their research on consumer products and to neglect products that are developed and manufactured as intermediate inputs into the production systems of other companies. This is perhaps unfortunate as the focus on designer or high fashion consumer products hides some of the complex ways in which design is used for commercial advantage in other sectors of the economy. This chapter has explored one such sector – the Norwegian maritime industry.

This chapter has shown the ways in which industrial design is much more than just a concern with consumer products. It shows how product value as represented through design is much more than just styling. It may, for example, be about developing improved process solutions that will enable a ship's operators to reduce the number of work-hours. Alternatively, it may be the design of equipment that enhances working conditions on board a vessel or offshore rig. In addition to functionality, these products must be able to meet the various public standards related, for instance, to quality, environmental considerations and safety.

Industrial designers target a technologically aware audience that wants something new; at the same time, the design intervention will be based on a collective understanding and experience (Farstad 2003). Higher-order business services such as industrial design targeting the maritime industry are characterized by a high level of competence within this sector. The design project also requires active participation by the client through numerous meetings and other ways of communication. Designs for products are delivered by internal teams or bought to the marketplace directly through project engagements or more indirectly through product deliveries in which the design process is organized by others. The independent designers identified within our study are companies with a generalist strategy in their degree of sector specialization (e.g. Hareide Designmill) and business sector specialists (e.g. LMG Marine and Vik-Sandvik), or they are functional specialists covering a specialized competence (Maritime Colours). Yet other design categories not included in detail in our studies include craft-based boat construction, which represents a geographical specialty, and silent designs, which are projects managed by other types of expertise, in this case, managers and craftsmen.

Most of these examples show the ways in which the design content reflects a competitive strategy providing a superior value. The result may be a product that is more efficient or more environmentally friendly through content, production process and use. It can also be the design of a product with a special function or cost advantage or a product that targets a specific use. Our research into industrial design within this sector shows that many of these advantages are important for design firms and their clients. Furthermore, we have found that client companies are mixing the management practice of standardization with product customization that offers flexibility in individual product details. Standards exemplified by the use of modular systems enable production to be spread geographically between low labour-cost locations that are close to major customers. The modular example also shows how concepts associated with, for example, Japanese production systems, are ways of organizing production across geographies that have spread to the car industry and then to other sectors such as the aerospace industry (Bryson and Rusten 2006), the furniture industry (Rusten and Bryson 2007) and, as shown in this study, the maritime sector. Finally, cases that are presented indicate the ways in which industrial design projects require expertise to cover a whole range of disciplines. This includes the capacity to organize, understand and communicate. Partnerships between individuals with different educational backgrounds and more efficiently trained industrial designers with a multitude of skills therefore are required.

Acknowledgement

I wish to thank LMG Marine, Ulstein Group, Hareide Designmill and Kongsberg Maritime for their information and illustrations for this research. Also thanks to Anne Kristin Wilhelmsen, SNF for editing.

References

Avisa Nordland (name of newspaper). 2009. 'Lover gassferger over Vestfjorden (Promising gas ferries over Vestfjorden)'. 17.02.2009.

Bergen Chamber 2005. 'Nye produkter i tradisjonsrik bransje (New products in a traditional industry)', *Newsletter Info* 1, 14.
Best, K. 2006. *Design management. Managing design strategy, process and implementation.* AVA Publishing, London.
Bryson, J.R. and Rusten, G. 2006. 'Spatial divisions of expertise and transnational "service" firms: Aerospace and management consultancy'. In Harrington, J.W. and Daniels, P.W. (eds) *Knowledge-based services, internationalization and regional development,* 79–100, Ashgate, Aldershot.
Coe, N.M., Kelly, P.F and Yeung, H.W.C. 2007. *Economic Geography. A contemporary introduction.* Blackwell Publishing, Hong Kong.
Dicken, P. 2007. *Global Shift. Mapping the changing contours of the world economy.* 5th Edition, Guilford, New York.
Farstad, P. 2003. *Industridesign.* Universitetsforlaget, Oslo.
Florida, R. 2002. 'The economic geography of talent', *Annals of the Association of American Geographers* 92, 743–55.
Fuchs, M. 2005. 'Internal networking in the globalising firm: The case of R&D allocation in German automobile component supply companies'. In Alvstam, C.G. and Schamp, E.W. (eds) *Linking industries across the world. Processes of global networking,* 127–46, Ashgate, Aldershot.
Karlsen, J.E., Quale, C., Rusten, G. and Jakobsen, S.E. 2005. 'Hvordan styrke gassregion Rogaland/Hordaland? (How do we strengthen the Rogaland/ Hordaland gas region?)'. *Report* 16/2005, Rogaland Research, Stavanger.
Kristensen, T. and Lojacono, G. 2002. 'Commissioning design: Evidence from the furniture industry', *Technology Analysis and Strategic Management* 14:1, 107–21.
Monö, R. 1997. *Design for product understanding,* Liber, Stockholm.
Montgomery, 2005. 'Beware "the Creative Class". Creativity and Wealth Creation Revisited', *Local Economy,* 20: 4, 337–343.
Nærings- og handelsdepartementet(Department of Trade and Industry). 2008. 'Et nyskapende og bærekraftig Norge (A creative and sustainable Norway)'. St.meld.nr.7 (2008–2009), Oslo.
Norwegian Design Council. 2005. 'Til topps med design i bunn (On top with design in the bottom)'. *Designåret 2005 Annonsebilag,* NDC, Oslo.
Norwegian Design Council. 2007. 'Maritime Møbler – rasjonalisering med nytt modulsystem (Maritime Møbler – rationalisation with a new module system)'. *Newsletter,* http://www.norskdesign.no, Published 11.10.2007.
Pratt, A.C. 2002. 'Hot jobs in cool places: The material cultures of new media product places: The case of South of the Market, San Francisco', *Information Communication and Society* 5: 1, 27–50.
Reve, T. and Jakobsen, E.W. 2001. *Et verdiskapende Norge.* Universitetsforlaget, Oslo.
Rusten, G. 2008. 'Designtjenester og geografi'. 243–265. In Isaksen, A, Karlsen, A and Sæter, B. (eds) *Innovasjoner i norske næringer. Et geografisk perspektiv,* 243–65, Fagbokforlaget, Bergen.
Rusten, G. and Bryson, J.R. 2007. 'The production and consumption of industrial design expertise by small- and medium-sized firms: Some evidence from Norway', *Geografiska Annaler* 89 B, 75–87.
Rusten, G. and Stensheim, I. 2007. 'Teknologiutvikling og design blant leverandører til oppdrettsnæringen. (Technology development and design

among suppliers to the fish farming industry)'. In Aarset, B. and Rusten, G. (eds) *Havbruk: Akvakultur på norsk*, 109–32, Fagbokforlaget, Bergen.

Scott, A. 2000. *The cultural economy of cities*. Sage, London.

Skipsrevyen. 2007. 'Gigantiske gassferjer overtar'. http://www.skipsrevyen.no 27.11.2007.

Statistics Norway 1994. 'Standard for næringsgruppering (Standard Industrial Classification)'. *Official Statistics of Norway*, Report C182, Oslo-Kongsvinger.

Statistics Norway 2008, Unpublished data.

Sweeney, G. 1996. 'Learning efficiency, technological change and economic progress', *International Journal Technology Management*, 11: 2, 5–27.

Ulstein. 2008. 'Newsletter from Ulstein'. 3, 1–12.

Web Resources

http://www.km.kongsberg.com, (accessed January 2009).
http://www.maritimt-forum.no (accessed January 2009).
http://www.ncemaritime.no/ (accessed January 2009).
http://www.norskdesign.no, (accessed January 2009).
http://www.mandal.umoe.no (accessed May 2009).
http://www.Vik-Sandvik.com (accessed April 2005).

10
Is Good Design Good Business?
Gorm Gabrielsen, Kjell Grønhaug, Lynn Kahle, Tore Kristensen, Thomas Plenborg and Ricky Wilke

Introduction

IBM CEO Tom Watson made the claim, that 'Good design is Good Business' at one of the first Aspen conferences on design in the 1950s (Watson and Petre, 1991). It seems generally accepted, that companies such as Apple, Alessi and SONY have gained competitive advantages from design (Dumas and Mintzberg, 1989; Kotler and Rath, 1988). Books like 'Winning by Design' (Walsh et al. 1992) and magazines like Business Week and newspapers such as the Financial Times have promoted the claim that investment in design as well as simply using good design enhances the competitiveness of firms. With a few notable exceptions (Blaich and Blaich, 1993; Gemser and Leenders 2001; Hertenstein, Platt and Veryzer, 2005), however, there is very limited evidence for such claims. In addition the concept of design is not well-defined and is often regarded as a superficial and superfluous fashion.

It seems important to investigate what exactly is meant by design, rather than ask open questions about what companies believe they invest in when they engage in such ill-defined activities. To take one claim, that investment in design rests on the ability to measure the investment precisely and to attribute costs precisely to design (Dirickx and Cool, 1989). This is difficult because while money paid to external consultants may be measured, it may be difficult to identify what was spent on product design, what was sent on communication design and what share was allocated to media expenditure. Accounting principles may lead to ambiguity, and much design is 'silent' (Gorb and Dumas 1987) and rarely accounted for.

The term 'good' requires further attention because the concept of 'design' is subject to ambiguity and the lack of an agreed definition.

While design is often thought of as easy to identify, there are numerous design specialisms: product-, communication-, web-, interface-, interior-, workplace-, transport-, light-, sound-, environmental-, furniture-, textile-, fashion-, automotive-, service-, experience-design to mention but a few professional specialisms found in phonebooks and web pages of design expertise. In such situations it is common to ask acclaimed experts in the field. This may require special attention, because expert knowledge may often rest on tacit knowledge and this makes it difficult to make explicit the articulation of tacit concepts.

This chapter investigates Thomas Watson's claim concerning the relationship between good design and good business. The key question is: why should design support corporate performance? The chapter is divided into four sections. The first section explores the definition and meaning of design and this is followed by the second section that locates design within a business context and reviews the body of knowledge. The third section explores the findings of an empirical study undertaken to add a more nuanced analysis of how different forms of design may contribute to improved performance in firms characterized by good design. The fourth section is a conclusion and discussion of the outcome and also considers priorities for further research.

Design

The design process and its outcome can be interpreted in many ways. Heskett (2002) underlines the importance of function and meaning when he notes that 'design, stripped to its essence, can be defined as the human capacity to shape and make our environment in ways without precedent in nature, to serve our needs and give meaning to our lives' (Heskett 2002, p. 9). This definition is broad and includes much of humanly made reality. For Heskett design must be characterized by its function, however, this is not sufficient by itself because design is also about meaning. Any individual designed object must be comprehensive and its purpose must be understood by the prospective user. Professional organizations that represent designers, for example, the Industrial Designers Society of America, stress both functionality but also the quality that the designer intends to convey through the design (IDSA, 2008) and in these terms design includes expressivity and credibility in that '...the professional service of creating and developing concepts and specifications that optimize the function, value and appearance of products and systems for the mutual benefit of both user and manufacturer' (Heskett 2002, p. 17).

An important distinction in design is reflected in the criteria developed for professional practice by the design professional bodies and also in the specifications developed to guide the formulation of training programmes that lead to educational qualifications in design. It is common in design curricula to distinguish between the product, logo and web design (Henderson and Cote, 1999). The oldest are product-, and graphic-, sometimes referred to as communication design and this reflects the rapid changes in the media. These are specialties taught in design schools and academies of art for over a hundred years and are often very close to the education of the arts. In the 20th century a separation was established in an increasing number of art schools that reflected a new perspective that situated design within a business context (Heskett 2005; Spivey 2005). While an increasing number of design services are based on research and conceptual work (Rosted et al. 2007), a significant number are based on the traditions taught in design- and art schools.

There seems to be no agreed standard for measuring the quality of design, but some acclaimed designers such as Dieter Rams of Braun have made an attempt to qualify what is meant by good design. Rams (1995) claims that good design must be 'innovative', 'useful', 'aesthetic', 'serves understanding the product', 'unobtrusive', 'honest', 'durable', 'consequent', 'environmentally friendly', and 'as little design as possible'. While these are general expressions, the terms are often used in connection with Danish design (Dickson 2006). Yet, the words used to characterize good design go no way towards the development of a scale that people can agree on. There is no single or even well-defined multiple scale that can be used to identify good design and many of the terms are still ambiguous.

Judgements of good design are often conducted in connection with contests and also in courts of law when dealing with design infringements. In such a situation experts are often consulted. Experts are people with considerable experience in dealing with complex and ambiguous issues, but all this still involves judgement (Sennett 2008). Even an expert's opinion may be difficult to express as it will be formulated around tacit knowledge (Polanyi 1958). The challenge is, therefore, to develop a research design that enables experts' knowledge to be identified and captured.

Design and business performance

This section briefly reviews past research regarding the relationship between design and business performance. We have made a distinction between assessments conducted by design promotion agencies and academic studies. The Design Council (Denmark) and similar organizations in other countries have been established as part of a national policy that is intended to promote the use of design. This policy is based on the general belief that good design is good business and that further applications of design will benefit the economy. As explained above, the design professions have many specialties and it seems very difficult for firms to identify the most suitable designer for a particular assignment. Therefore, many countries have established design centres to support the education of client firms and to try to demonstrate how design can be an asset in business. The design promoting organizations have found a need for setting the stage for design as a tool to increase the performance and profitability of firms. The absence of academic research on the relationship between design and competitiveness has meant that many Design Councils have had to commission management consultants to develop an evidence base.

An investigation conducted by the Danish Ministry of Commerce and Construction (Erhvervs- og Boligstyrelsen 2003) claimed that the turnover of firms using designers over a period of five years exceeded the turnover of companies not using design by 58 billion DKK. Moreover, companies that increased their purchase of design have a gross-income growth of 40 per cent more than those firms that incorporated less design in their products and services. The study is based on self-reported data gathered at one point in time. No information is given about the nature and quality of the design. 'Design' may include assistance from brand and communication consultants, which is priced very differently from product design, but we do not know this as the report lacks information on the way in which the data has been analyzed. For policy making this may be of little concern, but for research purposes it is a serious problem. Therefore it is difficult to judge the validity of the findings reported in this study.

The Design Council, London (2002 and 2006) has conducted a number of major studies, based on a methodology that is not explicitly reported in published form, but these studies do deploy statistical analysis. The data are based on self reported information from firms obtained from a questionnaire that requested respondents to provide information about their investment in design and the resultant financial outcomes. The overall conclusion is that those companies that use design perform much better than the average British firm. The design companies using more design launch more new products, their market share increases, they access more new markets and their turnover exceeds the industry average by a significant margin. Finally, these firms are less likely to compete on price, but rather on the products or services.

The Design Innovation Group of the University of Manchester and Open University (Potter et al. 1991) investigated the relationships between the payback time and the application of design and found that investments in the application of design had a fast payback compared to alternative investments made by firms. This study distinguished between products, packaging and graphic design; in addition, it was found that the latter had the fastest payback. However, generalizations of the findings are difficult, as the study relates to the use of a specific governmental support system, which provided firms with the 'free' use of a designer for a period of 45 days.

Gemser and Leenders (2001) in a study of design in the Netherlands explore the multiple meanings of design and what product design can actually do for a company. There is also a discussion on the advantages and challenges that result from the utilization of design. However, in their empirical analysis, Gemser and Leenders (2001) focus on expenditures used for professional design in product development in two industries, furniture and instrument manufacturing. The respondents were asked to rate their own performance vis-à-vis competing firms. Performance data were measured using a rank correlation. The hypothesis about a positive co-variation between design and performance was confirmed for the instrument industry but not for the furniture industry.

Hertenstein, Platt and Veryzer (2005) have reported a number of studies on the relationship between 'effective design' (industrial design) and business performance. Their investigation is based on an expert panel (the council of Design Management Institute) that was requested to rate designs from the best to the worst in nine industries. In the analysis they distinguished between 'effective design' and 'less effective design' and discriminated by a 'split half'. However, their study does not allow for an examination about the nature of design. The authors' report evidence for their hypotheses that the firms which are most effective at demonstrating good industrial design have a higher stock market return than those firms that have 'less effective design'.

The overview of past studies shown in Table 10.1 leaves us with many questions. For instance, are firms performing well because they decided to invest in design? Could it be that they were already performing at a high level prior to using design and now the use of design has just highlighted their existing strengths? Could it be the 'peacocks tail effect' that a company that performs well demonstrates this by a good design? This does not say whether design was the cause or the effect. It seems that in order to explore this issue that several years of firm performance data would have to be assembled and explored. This would be a complex process as design contains many different elements that

Table 10.1 Overview of design-performance studies

Authors	Type of research	Independent variable	Dependant variable	Methodology
The Design Innovation Group (1991)	Academic	Firms use a 'free designer' for 45 days	Payback times for product, packaging and graphic design	Based on a given sample, firms were asked about payback times for each kind of design
Gemser and Leenders (2001)	Academic	Firms' report of expenditures to pay designers	Firms' self-assessed rating of performance in furniture and instrument business	Regression of data from dependant and independent variables; correlation of design input vs. performance
Danish Ministry of Commerce and Construction (2003)	Consultancy report	Firms' report of investment in design	Firms' report of financial performance	Not available from the report
Platt, Hertenstein and Veryzer (2005)	Academic	Experts' judgment of total design for firms	Publicly available performance data	Regression based on split-half rankings of 'effective/less effective design'; correlation of design quality vs. performance
Design Council 2003, 2006	Consultancy report	Self-reported information from firms about their design	Financial performance measures	'Multivariate statistics'; no further qualification

are perhaps difficult to capture using financial data. We now turn our attention to exploring some of these dimensions.

Functionality

The functionality of a product is a major concern for the consumer. A product is acquired because a consumer expects a product to be able to function as a useful, effective thing and also perhaps a desirable object. A chair must be able to support a body in a stable position and a system designed to deliver medication must be able to deliver this effectively, accurately and easily. Convenience and ease of use are important criteria, as is the ability to apply the product with 'minimum force' (Sennett 2008).

Expressivity

Good design is defined according to meaning and context. A good design must be easy to understand. Does it look like what it is? A pair of scissors may look like a prototype scissors, but they should be easily identifiable as a pair of scissors but when a pair of scissors looks like something else it may be difficult for the consumer to identify them as a functioning object. Sometimes a subtle unobtrusive and simple design is the desired outcome. The chair '7' by the Danish designer Arne Jacobsen is a good example. It is used as one of the most common stacking chair in assembly halls, receptions, museums, private homes over the world. One reason is that this inconspicuous object fits into many environments, enabling it to make a strong statement compared to other objects.

Credibility

Finally, the criterion of credibility is a reflection that is similar to the feeling of value and the term 'reliability' appears to cover what the object of a specific design is intended to be. Credible means that the spectator easily comes to the conclusion that this is good design (Reber et al. 2004). An object must just look right; it should feel right without deliberation (Churchland 2002). The relation between the user and the object should feel 'seamless' and the user or consumer does not think of the thin line between the hand and the object. This becomes only evident if the object breaks down (Winograd and Flores, 1986). The feeling of seamlessness and value means a lot of cognition is going on, but it is tacit and itself unconscious.

A logo is intended for effective identification, typically of a company's brand. For the consumer, a logo means identification of the company delivering products and services (Orth and Malkewitz, 2008). In this sense, it is 'shorthand' for a number of design points which may include corporate architecture and even storytelling. The logo is a 'primer' for the expression of promises that a consumer should obtain if they purchase the good or service (Janiszewski and Meyvis 2001). The quality of the logo is dependent on its ability to signal a set of images and associations with limited levels of distortions, enabling the consumer awareness of the product and to raise the emotional attachment to the company. The logo does not need to reflect what a company is, but rather what it aims to become.

Web design concerns Internet interactions between the company and its stakeholders. Its functionality concerns effective communication and interactivity. A major issue is that stakeholders with a minimum of 'clicks' should be able to reach the required information and interact with the relevant departments in a firm. For this to happen the website must be reliable and fast. In addition, the design must be logical and expressive of the firms' intentions as well as the stakeholder's intuition of how it works. Easy access and a design that is accessible and simple to understand also contributes to a firm's credibility and this may be enhanced if the website design is also experienced as reliable.

The literature on the relationship between design and company performance leaves us with several unresolved questions. First, concerns about what is the function of design in business; what does the product-, logo-, and web design do for a company? Second, questions concerning how one might measure the quality of design for it to be identified as 'good' or 'bad' and, third, how can we measure firms' performance vis-à-vis design.

Design and business

Design is located at several positions vis-à-vis business functions. It is concerned with R&D, marketing, logistics and production in particular. A common distinction is made between product, communication and web design. One reason for this is that these three reflect different forms of expertise. Product designers and engineers have a different education compared to communication designers and their education is basically related to their technical and material engagement with products and technical solutions. Communication design is a particular expertise and specialization of design schools. Web design is the newest form of design specialization and it is taught both by design schools, communication designers and by computer departments. The connection with R&D is concerned with engineering principles and the materials used in the product. It concerns the product as well as the manufacturability of the product. A design that is developed by engineers or product designers must still meet the ten rules identified by Rams (1995) and these include issues such as usability, aesthetics, ensures that consumers will understand the product, unobtrusiveness, and so on. Marketing may be involved in understanding the market and qualifying the variations or standardizations required to create value in the marketplace that would result in the realization of profit. Marketing is also involved in communication decisions, for example, the development of a logo and packaging. Logistics and production is concerned with design in order to minimize use of precious materials and to find ecological and sustainable solutions and to deal with issues of standardization versus adaptation to target market niches.

A logo signals a pragmatic expression of a firms' intentions (Janiszewski and Meyvis 2001). When this article focuses on logos, it is meant as a simplification of communication design. A logo is used as an integral aspect of corporate communication both for internal organizational purposes such as corporate identity and signalling of values as it communicates these matters to other stakeholders (Hatch and Schultz 2003). The reason for this is that such a simplification is necessary as part of the research design that has been developed to support the analysis that is presented in this chapter.

The web design must be functional in the sense that it allows stakeholders, including consumers and users, to interact with a company easily and relatively cheaply. Companies that have reduced their overheads by introducing call centres may harm their reputations, the consumer experience of some call centres may be one in which a firm appears uninviting and perhaps arrogant. On the other hand, the cost of delivering a service via a call centre must be compared against the cost of delivering the same service via face-to-face interaction. A computerized response to frequently asked questions is one path that has been used by many companies. To work, the web design must offer convenience and fluency in operation as well as expressivity in reinforcing a firm's values. Stakeholders may be impatient and this implies that the credibility of a firm's interactive communication is one measure of a firm's perceived reliability.

Hypothesis

In this article design has been divided in three distinct elements, product, logo (communication) and web design. They represent distinctive design competencies and eventually affect business performance in various ways. These three elements are explored using three hypothesis.

Product

A good product design is positively correlated with a higher price and turnover. For a consumer to sacrifice their money to acquire the product it is usually evaluated both functionally and in terms of ease of performance. The product is usually evaluated by taking into consideration functionality, expressivity and credibility. The argument for this hypothesis is that enhanced consumer value is due to the higher quality of the design and this leads to a higher price and increased sales. This in turn leads to better business performance. This produces the first hypothesis:

H1: The quality of product design is positively co-varied with financial performance.

Logo

The logo is part of a firm's identity assets and a logo is given particular attention and is constantly modernized to signal other changes in a company. In markets, the logo is a major source of identification and a tool for building a consistent image. This involves brand communication, where the assumption is that marketing communications may influence consumers. The logo may attract the awareness of the user, and yet the decision to purchase may be driven by other elements.

A new logo may be used to signal a new direction and new values. A logo may become a link between the company and the customer supporting remembrance and creating loyalty. An effective logo should, therefore, increase turnover. If the logo is new, the valuation (Janiszewsky and Mayvis 2001) may be negative because recognition takes time and credibility may be missing. In such instances a negative valuation should be expected. The consumer does not recognize the logo as a credible expression of what the company stands for. Only a lot of additional communication may influence consumers to understand that the company is trying to alter its image. In such situations, a negative correlation between product and logo may be expected. This gives rise to the second hypothesis:

H2: That the logo design co-varies with a firm's financial performance.

Web

Web design is essential for interactivity between a company and its customers. Online services and support is an important issue where users expect to obtain an immediate response. The third hypothesis concerns the relationship between a firm's web design and the way in which this influences interactivity and the way in which services are provided to users. This provides benefits in terms of information efficiency and network externalities to the business. A good web design attracts new users and may increase consumer transactions with a company. There should, therefore, be a positive correlation between good web design and performance. This produces the third hypothesis:

H3: A web design is positively co-varied with financial performance.

Sample and research design

To investigate these questions a study was conducted using a random sample of 25 of the 100 largest Danish companies. The primary aim was to explore the relationship between good design and good business performance. This section describes the research design by exploring the sample, the financial data and the design data.

Sample

The sample of firms is taken from a list of the '100 largest Danish firms' that is produced by the Danish newspaper Børsen. The data were corroborated by public databases for the period 2001 to 2005. The companies included in this analysis were randomly selected from the list by identifying every third company. Unfortunately, for some companies essential information was unavailable and these firms could not be used in the analysis. Some companies were subsidiaries of larger companies or local sales departments of multinational firms. Others were publicly owned and finally some did not have a legal form that required them to deposit a publicly assessable copy of their accounts. The remaining 25 companies are dominant local players operating in industries such as food, transportation, pharmaceutical, manufacturing and energy. A design dominant industry such as the furniture industry was not included in the sample.

Financial data

Penman (2007) regards ROI (Return on Investment) as the superior performance measure. We define ROI as operating profit before tax divided by total assets. PM (Profit Margin) is defined as operating profit before tax divided by net turnover. ROE (Return on Equity) is defined as net earnings divided by book value of equity. In addition, we apply two market based profitability measures; growth in market value of equity and market to book value. For example, Penman (2007) finds that a high market to book ratio indicates a high future level of profitability. We measure growth as growth in turnover, net assets, ROI and PM, respectively. As a proxy for size we apply net turnover, total assets and the market value of equity, respectively. These proxies were available for the 25 firms over the period of five years from 2001 to 2005.

Accounting data imposes challenges and fluctuates from year-to-year due to a variety of reasons such as investments, mergers and acquisitions, changes in accounting principles, divestment of assets; many of these events have little to do with the basic value creation or design. To reduce this problem we summarize the five numbers in two ways. As a measure of financial values, we calculate the mean over the five years 2001–2005 and as a measure of growth we calculate the mean growth rate over the five years. The reason for using average and growth rates is that many different influences affect performance in a single year, such as business cycles, depreciations, merger and acquisitions. The performance measures will be crude measures, but by taking averages and growth rates it is possible to adjust for some fluctuations that are not due to design. To take averages reduces the variance and enables a smaller scope for co-variation. Therefore, the co-variation found should reflect a stronger connection between the independent and the dependant variable. The risk however, as will be clear is that no co-variation may be found. The financial data are summarized in the Appendix (Table A.1).

Design data

To capture the meaning of 'good design', we used an expert panel. The Danish Design Centre and the Association of Danish Designers were asked to select design experts. Each of them was asked to select five designers with a reputation for being among the best designers and the best referees in design contests and legal settings. They all had extensive industrial experience. Out of the total of ten that were identified some were unavailable at the time of the experiment, but seven design experts participated in the expert panel. Each design expert was presented with a screen that contained two images of designed products that enabled them to undertake a paired comparison (Figure 10.1).

The illustrations were the experimental stimuli. They were shown in a random order to avoid a fatigue effect. This technique is a 'paired comparison design' method that is often used in sensory experiments (Gabrielsen 2000, 2001). The assessment of a design by the experts was made by moving the cursor on all three scales from the middle position to either extremes or to some position depending on the each expert's assessment of a design. A modest discrimination would be reflected in a moderate position between the two illustrations. There were three scales and each expert was asked to assess functionality, expression and credibility. An interview was conducted to investigate what the expert understood by these terms. The experts were asked only to move the cursor and then the location was measured. By avoiding any reference to numbers we attempted to remove some biases, for example, those of a numeric kind like regression to the mean. The reason for paired comparisons is that it yields a stronger preference statement compared to asking an expert to rate stimuli for an image and then calculating

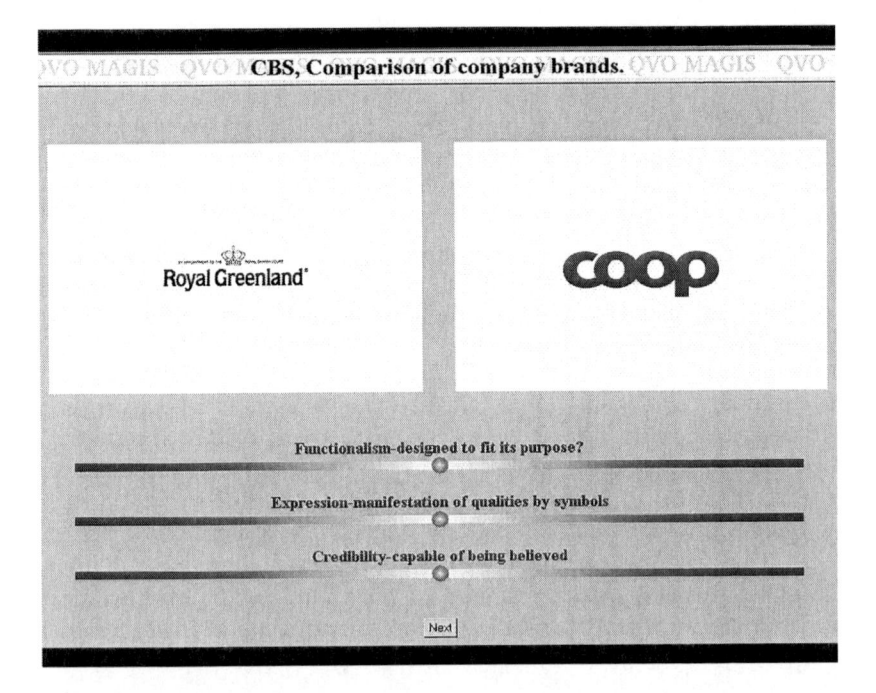

Figure 10.1 Logos from Royal Greenland and COOP compared with respect to 'Functionality', 'Credibility' and 'Expressivity'

differences. This is similar to asking a jury to state whether a person is regarded as guilty or not guilty as compared to giving a reason for it. It is a difficult cognitive operation and the assumption is that the simple paired comparison is more valid.

Samples of design were selected for each of the 25 firms and these were downloaded from company websites and set up as a paired comparison in a set of experiments with three parts; one part comparing products, the second part comparing logos and the third part comparing web designs.

The absence of actual physical designs may be partly compensated by the fact that the experts are all experienced and have all been exposed to the designs of the large companies. To prevent problems of validity in judgement, all experts were asked to indicate if they were personally involved with any of the firms. That was not the case.

Each expert was placed at a computer screen showing two illustrations of products, one illustration to the left and one illustration to the

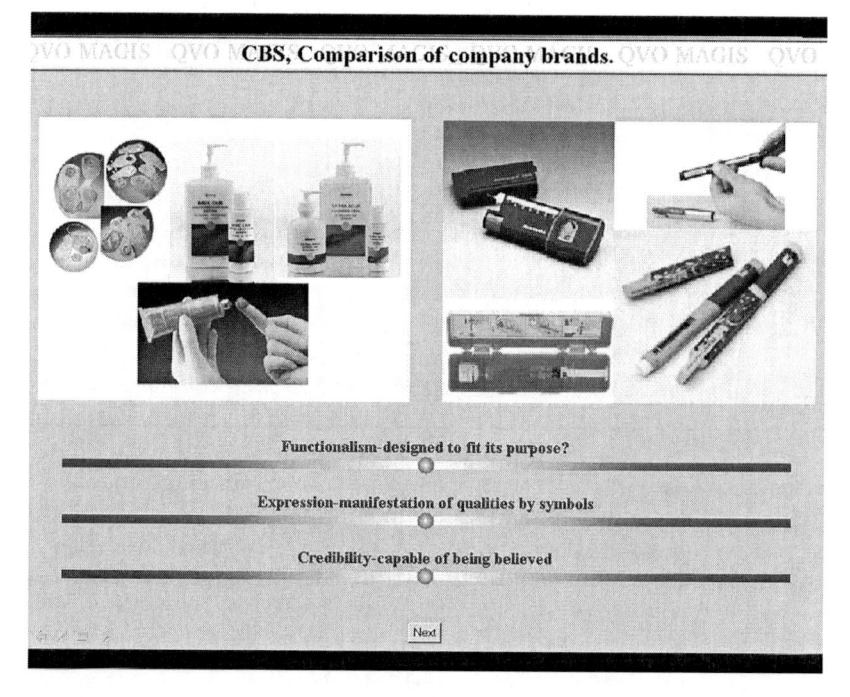

Figure 10.2 Products from Nomeco and Novo Nordisk compared with respect to 'Functionality', 'Credibility' and 'Expressivity'

right. To give a better impression of the products a projector was used to enlarge the design (Figure 10.1 and Figure 10.2).

The ratings were recorded electronically by the expert moving a cursor from the middle to the left or to the right, indicating which was the favoured design. A rating close to the middle indicates that no strong preference exists and the two designs were considered equal. A rating close to the right end of the scale indicates a high preference for the right illustration compared to the left illustration – and vice versa. After the preferences were rated, the expert pressed a button marked 'next' and a new screen with two new illustrations of products appeared and the scores were registered electronically.

Some training was provided to ensure that the experts understood the process and some trial comparisons were performed to ensure that they understood the procedure. After fulfilling the first part of the experiment, comparing products, the expert continued to the second part, comparing logos, and finally to the third part and the comparison of web designs. In each part of the experiment the pair of illustrations was presented in a random order. To reduce the demands on the designer the number of comparisons in each part of the experiment was reduced to 25 comparisons following an experimental design allowing for preference scales to be estimated. The time to value and respond to the designs was approximately one hour per expert.

Statistical analysis of comparisons

At the end of each part of the experiment ratings were converted into numbers by the researchers. The distance from the middle of the scales to the noted mark was measured (electronically), positively to the right and negatively to the left. For the quality q (q=1 ~ functionality, q=2 ~ Credibility and q=3 ~ Expressivity) the corresponding value of the score of expert p with firm i to the left and firm j to the right is denoted y_{ijqp}. In any event it was assumed that the numerical scores increase with the strength of the preference for i over j, and that equal but opposite preferences corresponds to equal but opposite scores. If, for example, y_{ijqp} is positive j is preferred over i, if y_{ijqp} is negative i is preferred over j. The underlying assumptions of the mathematical model are that all the y_{ijqp} are independent random variables and normally distributed with mean value $E(y_{ijqp}) = \xi_{ijqp}$ and the same variance, σ^2.

The purpose of the paired comparison is to use each of the qualities to arrange the firms along a rating scale and this scale represents the assessed preferences of two firms that can be attributed to the difference of the two firms on a continuous scale. The usual way of expressing a rating scale is that there exists a correspondence to the firms so that $\xi_{ijqp} = \alpha_{jq} - \alpha_{iq}$. As only the differences of α's are of interest one can add the restriction that for each quality the α's should sum to zero. The model is a general linear model and the estimation of α's, therefore, is straight forward.

The normality assumption is unlikely to be a problem, due to the robustness of ANOVA. Scale biases due to a higher density at the endpoints of the scale may cause problems; however, subsequent analysis of the raw data showed no 'jamming' at the endpoints. Also, probit-diagrams of standardized residuals showed no irregularities. It was not possible to obtain all three qualities of design and they were merged into one concept, good design.

Descriptive statistics

The arrangement of the 25 firms into a rating scale for each of the three qualities is shown in Table A.2 in the Appendix.

The relation between design quality and financial performance

The data generated are shown in Table A.2. The vertical axes show the correlation coefficients for product, logo and web design respectively.

Table 10.2 Rank correlation matrix, Spearman's rho

Size	N =	Product	Logo	Web
Net turnover, mean 5 years	25	0.433*	0.136	0.35
Net assets, mean	25	−0.085	−0.497*	−0.019
Market value of equity, mean value	13	−0.456	−0.258	−0.06
Growth				
Net turnover growth rate over 5 years	25	0.182	−0.343	−0.106
Net assets growth	25	0.171	0.161	−0.061
Return on investment[a], growth rate	25	0.205	0.123	0.08
Profit margin[b], growth rate	25	0.442*	0.145	0.188
Profitability				
Return on investment[a], mean 5 years	25	0.07	−0.398*	0.063
Profit margin[b], 5 year mean	25	0.104	−0.313	0.209
Stock market equity growth rate, 5 years	13	−0.379	0.22	−0.363
Market to book, mean	13	−0,104	0,555*	0,39

*Correlation is significant at the 0.05 level (two-tailed).
[a] Return on investment is defined as operating profit before tax divided by total assets.
[b] Profit margin is defined as operating profit before tax divided by net turnover.

The horizontal axes show values for various performances. The correlations are between the rankings of three forms of designs and a variety of performance measures. The actual measures are seen in Table 10.2.

The rank correlation between net turnover and product design is positive and significant at the 5 per cent level. Furthermore, net assets are negatively correlated with logo design at the 5 per cent level. As Table 10.2 shows, the signs in the correlation matrix differ; they are positive for product and negative for logo. Again, this is seen as a sign, that the nature of product design and the other forms follow different rules. It is quite possible for product design to be positive and graphic design to be negative. This is because the product is what is offered to the user or consumer at the sacrifice of price, while graphic design may communicate what the firm aspires to become, but has not yet achieved. This finding also strengthens the claim that the relation between product and other forms of design is dominated by product design and not vice versa.

Growth only explains the quality of the design in a modest way. The correlation coefficient between growth in profit margin and product design is positive and significant at the 5 per cent level. Other growth related correlation coefficients remain insignificant. The quality of web design is neither explained by size, growth and profitability.

Analysis and discussion

There are positive correlations between product design and the 'net turnover mean over 5 years' and 'profit margin, growth rate' and negative correlations for logo and 'net assets, mean', 'return on investment mean 5 years' and 'market to book, mean' (Table 10.2). Evidently this is a weaker outcome than a strong correlation for ROI. Still it shows that there is a correlation. The expectations based on the literature were that we would find a positive relation between all three forms of good design and business performance. What can be a more serious challenge is that firms have different strategic contexts and different histories and rhythms. Also we have no data to ascertain whether some companies radically changed their designs within the last five years. Specific insights will give a much more nuanced understanding in future studies. H1 was therefore accepted, H2 accepted and H3 hypothesis was rejected.

The findings were correlations and these are weaker statements than causality (Cook and Campbell 1979). The correlations were chosen instead. With the level of complexity involved in the combined valuation of a large number of designs combined with performance measures of companies' performance, it seems impossible to hope for more. Yet inclusion of some variables for strategy and localization of design may give deeper insights.

The findings indicate that the main driver of company performance is product design. Logo- and web design may follow product design, but there is nothing to indicate that a good logo and/or web design can compensate for bad product design. The outcomes may be a direct reflection of the fact that the three forms of design were organized in different ways in the firms. It is evident, that a logo is a gross simplification of corporate communication. Yet the literature indicates that a logo is a vital asset for a company.

Are there good examples of the dominance of product design in current business? There are probably both good and bad examples. Yet often they are mixtures. The Novopen© with its many developments to support people with diabetes and subsequently the delivery of growth hormones to children may serve as an example. The pen was developed over many years and was always focussed on user needs. A classic example may be the Volkswagen. Today it may be considered as a major brand, but the current performance of the company is not due to its logo, but rather persistence in technology and product design. A negative example is the firm of Bang & Olufsen which has lost a major share of its capital for ignoring the fact that consumers no longer want to purchase large and dominant TV-screens for their homes when they can purchase cheaper and more inconspicuous screens. No communication campaign or the development of shop concepts seems to have compensated them for this shift in consumer preferences. Other examples are ambiguous, for instance Apple, who's classic Mac was very much a strong product design and brand. Apple's success is based not only on the development of a superior product design, but also on the creation of a series of successive products and the creation of a cult status around the brand.

Conclusion

This chapter has explored the distinction between product design versus logo design and the financial performance of a firm. A research design has been created and tested that is a development to the methodologies deployed in other studies. Our approach is based around a more nuanced methodology constructed around the use of expert panels and the application of paired comparisons to create a 'realistic setting'. This technique enabled us to refine the classification of products in to those that reflected 'good design' and those that did not. We argue that this technique is a useful development that has the ability to explore the relationship between firm performance and good and bad design. It seems that the future may show much more ambiguity and a lack of clear-cut examples of one form of design in comparison to another. Yet, it seems that not even Apple can avoid the problems that come from the development of a poorly designed product. Further research may substantiate this. Web design was not found to be a significant factor in relation to firm performance in this study. That does not mean web design is not important for other firms. A study including e-commerce and even iPod or Amazon.com might reveal a different perspective on this issue.

In our opinion based on the evidence presented in this chapter it is safe to claim that Tom Watson's words are true; companies may use design to improve their business. Yet there are more unanswered questions. Future studies of design in business must deal with all these issues and the challenges are complex. Research in to the relationship between design and corporate performance has been rather limited in scope. The extant literature has contributed to a common body of knowledge, but more comprehensive and sophisticated research designs must be developed to address Tom Watson's claim. What is required is further quantitative and qualitative studies. 'Qualitative studies', for example, case studies, could provide valuable insights since they have the potential to develop an intensive analysis of value creation, design and company performance. Case studies have the potential to identify issues, processes and strategies that could unlock some of the strategic considerations that lie behind the relationship between good design and good business.

Appendix

Table 10.A.1 Financial data

Proxy	N	Mean	Std. Deviation
Net turnover, mean[a]	25	13516.896	12819.026
Net turnover, growth rate[b]	25	0,092	0,247
Profit margin (PM)[c], mean	25	0,063	0,070
Profit margin (PM), growth rate	25	–0,004	0,016
Return on investments (ROI)[d], mean	25	0,060	0,060

Continued

Table 10.A.1 Continued

Proxy	N	Mean	Std. Deviation
Return on investments (ROI) growth rate	25	−0,005	0,019
Return on equity (ROE)[e], mean	25	0,088	0,150
Return on equity (ROE) growth rate	25	−0,022	0,066
Growth in stock price	12	4.510	45.920
Market to book[f], mean	13	2,715	2,764

[a] Mean is mean of the years 2000 to 2004.
[b] Growth rate is the mean growth rate in the years 2000 to 2004.
[c] Profit margin (PM) is defined as operating profit before tax divided by net turnover.
[d] Return on investments (ROI) is defined as operating profit before tax divided by total assets.
[e] Return on equity (ROE) is defined as net income divided by book value of equity.
[f] Market to book is defined as market value of equity divided by book value of equity.

Table 10.A.2 Rating scales for the three qualities

Firm	No.	Logo	Product	Web Design
Danisco A/S	1	0,76	0,47	0,96
Danish Crown AmbA	2	−0,59	−0,70	−0,81
Arla Foods amba	3	0,35	0,21	0,81
DLG Service A/S	4	0,52	0,41	−0,13
Royal Greenland Seaf	5	1,05	−0,43	−0,66
Coop Danmark A/S	6	−0,69	−0,16	0,04
DFDS Seaways A/S	7	0,51	0,63	−1,97
SAS	8	2,01	2,19	0,85
DSB A/S	9	0,32	1,04	1,86
Sophus Berendsen A/S	10	−0,69	−0,16	0,04
H, Lundbeck A/S	11	−1,08	0,74	1,59
Coloplast A/S	12	−0,64	0,11	0,02
Novo Nordisk A/S	13	−0,96	1,77	1,16
Nomeco A/S	14	−0,69	−0,16	0,04
FLS Industries A/S	15	0,40	−0,47	−0,15
Rockwool A/S	16	2,43	1,28	0,15
Aktieselskabet Nordi	17	0,52	−0,57	−1,34
MT Højgaard A/S	18	1,06	0,52	1,21
Flextronics Internat	19	−1,32	−0,57	−1,51
Danfoss A/S	20	−0,69	−0,16	0,04
NESA A/S	21	0,59	−1,34	0,09
Aarhus Oliefabrik A/S	22	−0,64	−2,14	0,04
Vestas Wind Systems	23	−1,82	−2,12	−1,85
Kuwait Petroleum	24	0,18	0,12	0,53
NEG Micon A/S	25	−0,20	−0,38	−1,03

References

Blaich, R. and Blaich, J. 1993. *Product design and corporate strategy: Managing the connection for competitive advantage.* McGraw Hill, New York, NY.

Churchland, P. Smith 2002. *Brain-wise studies in neurophilosophy.* A Bradford Book MIT Press, Cambridge, MA.

Cook, T.D. and Campbell, D.T. 1979. *Quasi-experimentation design & analysis issues for field settings.* Mass Houghton Mifflin Company, Boston, MA.

Design Council 2002. *Competitive advantage through design.* (http://www.karo.com/portfolio/images/ideaspdf/Competitive%20Advantage%20Through%20Design.pdf) (accessed May 2007).

Design Council 2006. *Design in Britain 2005–06.* (http://www.designfactfinder.co.uk/design-council/pdf/DesignInBritain200506.pdf) (accessed June 2007).

Dickson, T. 2006. *Dansk Design.* Gyldendal, København.

Dirickx, I. and Cool, K. 1989. Asset stock accumulation and sustainability of competitive advantage, *Management Science* 35:12, 1504–11.

Dumas, A. and Mintzberg, H. 1989. Managing design: Designing management, *Design Management Journal* 1:1, 37–43.

Erhvervs- og Boligstyrelsen (Danish Ministry of Commerce and Construction) 2003. *Designs Økonomiske Effekter (Economic effects of design)* (http://www.ebst.dk/file/1638/designeffekter.pdf) (accessed June 2007).

Gabrielsen, G. 2000. Paired comparisons and designed experiments, *Food Quality and Preferences* 11, 55–61.

Gabrielsen, G. 2001. A multi-level model for preferences, *Food Quality and Preferences* 12, 337–44.

Gemser, G. and Leenders, M. 2001. How integrating industrial design in the product development process impacts on company performance, *Journal of Product Innovation Management* 18:1, 28–38.

Gorb, P. and Dumas, A. 1987. Silent design, *Design Studies* 8:3, 150–56.

Hatch, M.J. and Schultz, M. 2003. Bringing the corporation into corporate branding, *European Journal of Marketing* 37:7–8, 1041–64.

Henderson, P.W. and Cote, J.A. 1999. Guidelines for selecting and modifying logos, *Journal of Marketing* 62 Winter, 14–30.

Hertenstein, J., Platt, M. and Veryzer, R. 2005. The impact of industrial design effectiveness on corporate financial performance, *Journal of Product innovation Management* 22, 3–21.

Heskett, J. 2002. *Logos and toothpicks.* Oxford University Press, Oxford.

Heskett, J. 2005. *Shaping the future design for Hong Kong. A strategic review of design education and practice design task force*, The Hong Kong Polytechnical University.

Industrial Designers Society of America (IDSA). 2008. ID Defined, Downloaded from (http://www.idsa.org/absolutenm/templates/?a=89&z=23) (accessed October 2008)., **IDSA:** Dulles, VA

Janiszewski, C. and Meyvis, T. 2001. Effect of brand logo complexity, repetition and spacing on processing fluency and judgement, *Journal of Consumer Research* 28, 18–32.

Kotler, P. and Rath, G. 1988. Design: a powerful but neglected tool, *The Journal of Business Strategy* Autumn, 16–21.

Orth, U. and Malkewitz, K. 2008. Holistic package design and consumer brand impressions, *Journal of Marketing* 72, 1–18.

Penman, S. 2007. *Financial statement analysis and security valuation.* 3rd edition, Mcgraw-Hill, New York, NY.

Polanyi, M. (1958). *Personal knowledge: towards a post-critical philosophy.* Chicago, IL: University of Chicago Press.

Potter, S., Roy, R., Capon, C., Bruce, M., Walsh, V. and Lewis, J. 1991. *The benefits and costs of investments in design: Using professional design expertise in products engineering and graphics*, Report of the Design Innovation Group, Manchester.

Rams, D. 1995. *Weniger, aber besser less but better.* Jo Klatt Design+Design Verlag, Hamburg.

Reber, R., Schwarz, N. and Winkielman, P. 2004. Processing fluency and aestethic pleasure: Is beauty in the perceivers processing experience? *Review of Personality and Social Psychology* 8:4, 364–82.

Rosted, J., Lau T., Høgenhaven C., Johansen P. (2007), *Concept Design, How to solve complex challenges of our time.* FORA: Copenhagen

Rumelt, R. 1984. *Towards a strategic theory of the firm.* In Lamb, R. (ed.) *Competitive strategic management.* Prentice-Hall, Englewood Cliffs, New Jersey, NJ.

Sennett, R. (2008) *The Craftsman.* New Haven. CT: Yale University Press.

Spivey, N. 2005. *How art made the world a journey to the origins of human creativity.* Basic Books, New York, NY.

Walsh, V., Roy, R., Bruce, M., Potter, S. (1992), *Winning by Design: Technology, Product Design and International Competitiveness*, Blackwell, London.

Watson, T. Jr and Petre, P. 1991. *Father and Son Inc. My life at IBM and beyond.* Bantam Books, New York, NY.

Winograd, T. and Flores, F. 1986. *Understanding computers and cognition: A new foundation for design.* Ablex, Norwood, NJ.

Index

Aalto, Alvar 54, 63, 69
Aalto University 57, 60
Adorno, Theodor 61
advertising 28, 31, 106, 142
Agreement on Textiles and Clothing (ATC) 99
Albert (prince) 36
apparel industry, in Montreal 94–114
Apple iPod 117
Apple Macintosh computers 8, 237
Arabia 69
Arabianranta 67–73, 75
architecture 70, 142
Art and Design City Helsinki (ADC) 68
artisanal labour 143, 144
arts 60
  design and 40
  manufacturing and 37–8
arts and crafts movement 178
AutoChief C 20 210, 211
automatic washing machines 2–3

Bang & Olufsen 237
Bangalore, India 64–5
Barbara K 180–1
Barletta, Martha 187
Bauhaus 178
Better by Design programme 52
Blåkläder 181–3
body types 179–80
border crossing 180–1, 184
Boulton, Mathew 33–5
branding 9, 106, 109, 120, 142, 170–1
Britain Can Make It (BCMI) Exhibition 40, 43, 45
British design 38–40
British Empire Exhibition 43
Buffalo, NY 85
business, design as good 220–39

business performance, design and 222–7
Butler, Judith 174

Calgary 126
California 82, 85–6
Camp, Alan 156, 158
Canada
  design work in 117–37
  fashion industry 93–114
capitalism 30–1, 54, 175
  economic geography and 55
  factory 29
  performative 56
  soft 57
car design 186–90
catalogue selling 26–7
Central Business Districts (CBDs) 147, 148, 159–60
Central Register of Establishments and Enterprises (CRE) 202
children's clothing 178–9
China 1
  low value-added goods in 1
  offshoring to 128–9
Chinese Chippendale 29
Chippendale, Thomas 29
cities
  creative communities in 199
  cultural economy and 141–4
  geography of design and 141–64
  interdependencies in 144–5
  London 150–8
  property markets in 148, 150
  resurgent 141–4
  U.S. industrial design firms concentrated in 82–6
  Vancouver 158–62
City Fringe (London) 151–8
class conceptions 59–60
Clerkenwell, London 154–6
clusters 24–5
Coca-Cola bottle 7–8

243

244 *Index*

cognitive cultural economy 145
Cole, Henry 41–2
colonization 28–9
Commercial Committee 35
communication design 227
competition
  based on intangibles 1–2
  bases of 5, 25
  design-based 1–3, 7, 10, 28–31, 33–5
  firm-based 31
  foreign 28–9, 46
  globalization and 1
  Industrial Revolution and enhanced 28–31
  non-price-based 6–7
  in Norwegian maritime industry 195–217
  price-based 1, 6, 29–30
competitive advantage 31–2
  design-based 10
  place-based 6
competitiveness 55–6
  corporate 3, 5–10, 31–2, 170–1
  creativity and 117–18
  industrial design and 195–217
  international 57
conceptual design 183–6
consumer behaviour, national 3
consumer electronics 120
consumer needs 4
consumer recognition 9
consumers
  distant 97–8
  female 175–7, 186–90
  local 105–6
consumption, production and 95–6
contextual knowledge 95–7
continuous innovation 29–30
contract workers 129–31
copyright protection 37
corporate branding 142
corporate competitiveness 170–1
  design and 5–10, 31
  industrial design and 3, 31–2
corporate form 25, 46
Council for Industrial Design (COID) 40, 43, 45
craft, design and 37–8

craft production 26, 31
crash-dolls 179–80
Creative Cities Network 52
creative city movement 145
creative class 51, 53, 58–60, 65–7, 74, 198–9
  values of 65–7, 68, 72–3
creative communities 198
creative economy 13, 50–75
creative industries 60, 61, 121
Creative London 52
Creative Tacoma 52
creative workers 58
creativity 6
  commodification of 59–60
  competiveness and 117–18
  in contemporary urban economy 119–22
  defined 95
  technology and 60–1
  values of 65–7
creativity discourse 51–62, 73–4
cultural dimension, of design 12
cultural economy 141–4, 145, 156
cultural industries 60, 61, 62, 121
cultural intermediaries, role of 94–114
cultural production 150
  globalization and 97–8
  spatiality of 143–4
cultural products 142
cultural urbanism 145–9
culture 52, 60, 61
Curtis, Alistair 64–5

d'Aveze, Marquiis 40
Delfina Trust 158
demand-size perspectives, on U.S. industrial design firms 89–90
design
  art and 40
  British 38–40
  business and 220–1, 227–39
  business performance and 222–7
  car 186–90
  communication 227
  conceptual 183–6
  in contemporary urban economy 119–22

design – *continued*
  as contextual process 95–8
  copying 35
  corporate competitiveness and 5–9, 10
  craft and 37–8
  cultural dimension of 12
  defined 4–5
  effective use of 120
  fashion 93–114
  firm-based competition and 31, 33, 46
  gender and 169–92
  geographical dimension of 12
  geography of 2–3, 122–7, 141–64, 198–206
  importance of 2–3
  incorporation in value chain 5
  inimitability and 5–9
  interdependencies 144–5
  investment in 220
  in London 150–8
  neutral 179
  politics and 178
  product 227–9, 236
  quality of 221–2, 231–6
  role of 9–13, 120
  spatial model of 145–9
  universal 11, 178
  as value added 93, 120
  in Vancouver 158–62
  web 227, 228, 230, 237
  see also industrial design
design-based competition 1–3, 7, 10
  Industrial Revolution and 28–31
  Soho Works and 33–5
Design Council 222, 223
design data 231–4
design education 25, 214–16, 222
designer products 3, 9–10
designers 10–11, 58–9
  in Canada 122–7
  independent fashion 94–114
  project-based work by 129–33
  research process by 12
  role of 30–1
  types of 11
  in U.S. 127
design exhibitions 40–5, 47

design firms
  degree of specialization 204–5
  industrial, in U.S. 81–91
  small 202–4
design industry
  in Finland 50–75
  firm practices in 129–33
  spatiality of, in cities 145–9
  in U.S. 81–91
Design Innovation Group 223–4
design inspirations 33, 35
design pillar 30–1
design process 11, 221
  client involvement in 205–6
design profession, development of 38, 46
design registration 11
design schools 39, 222, 227
design teams 10–11
Design Tradeoff Analysis (DTA) 82
design work
  localized nature of 129–33
  offshoring of 127–9
  project-based, in Toronto 117–37
  proximity issues and 133–4
  removed from production 118–19
deskilling 29
Detroit 85
developed economies, high valued-added goods and 1–2
differentiation 29, 32
diversity 67
division of labour 26–31, 117–18, 163
Dove's Evolution 170
Duel 186

East India Company 29
Eastman Kodak 29
economic policy
  design exhibitions and 40–5
  incorporation of industrial design into 35–40
economy
  creative 13, 50–75
  cultural 141–4, 145, 156
  knowledge 55, 144
  national 23–4
  New 142, 144

economy – *continued*
  service   143
  urban   119–22
education, design   25, 214–16, 222
Ehrnberger, Karin   184
employment, in Canadian design sector   122–7
engineers   10, 12, 30, 31
English watch industry   27–8
Ergon Grips   180
Etruria porcelain factory   26–7
exhibitions   40–5, 47
expositions universelles   43
expressivity   226

factory capitalism   29
factory production   26
fashion designers   11, 104–6
fashion industry
  cultural intermediaries in   106–12
  Montreal   93–114
fashion magazines   105, 106, 113
fast fashion   93
female bodies   179
female car designers   170
female consumers   175–7, 186–90
feminism   171, 175, 191
Festival of Britain   45
financial data   230–1, 238
fine arts   37, 61
Finland
  creative economy in   50–75
  design sector in   62–7
  national design programme   52, 57
Finnish Design   50–1, 63–7
firm-based competition   31, 33, 46
Florence   143
Florida, Richard   51, 59, 65, 67, 145
Fordist production   143
foreign competition   28–9, 46
France, design exhibitions in   40–1
Franck, Kaj   54
functionality   226
furniture designers   11

gender
  commercialization of   172–7
  as cultural construction   173–4
  design and   169–92
  performativity of   174
gender awareness   169
gendered design practice   177–83
gender theory   169–70, 171
Geneva   143
geographical context, of U.S. industrial design firms   82–6
geography
  of design   2–3, 122–7, 141–64, 198–206
  of maritime sector   206–7
  of production systems   199–201
George IV   37, 41
global cities   55
  see also cities
globalization   55, 170
  competition and   1
  cultural production and   97–8
global manufacturing, dynamics of   2
government policy   13, 23–5
  design exhibitions and   40–5
  incorporation of industrial design into   35–40
Great Depression   32
Great Exhibition of 1851   37, 39–40, 42–3

handmade products   26
Hareide, Einar   12
Helsinki   67–73, 143
Helsinki Virtual Village (HVV)   71–2, 75
high valued-added goods   1
Hirsch, Paul   95–6
hollow-ware industry   32–3
Horkheimer, Max   61

ICT sector   61, 62, 63
IKEA   11
imports   28–9
inclusiveness   169
independent stores   102–6
India   1, 64–5
industrial clusters   24–5
industrial design   142
  corporate competitiveness and   3, 5–9

industrial design – *continued*
  creative economy and 52
  defined 4–5
  vs. engineering 197
  in Finland 52–3
  identification of 201–2
  impact of 3
  incorporation of, into national economic policy 35–40
  Industrial Revolution and enhanced 31–5
  national competitiveness and 46–7
  in Norwegian maritime industry 195–217
  role of 9–13
  see also design
industrial designers 30–1, 58–9
industrial design firms, in U.S. 81–91
industrial design policy, design exhibitions and 40–5
industrial districts 55, 143
industrial exhibitions 36–7, 40–1
industrial gentrification 150
industrial parks 149
Industrial Revolution 12–13, 26, 143
  division of labour during 26–31
  enhanced competition and 28–31
  industrial design and 31–5
  price-based competition and 29–30
Ingold, Pierre Frédéric 28
inimitability 5–9
inner city 147–8, 163–4
innovation 29–30, 32–3, 119–22, 144–5
innovative milieu 55
intangibles
  competition based on 1–2
  investment in 2, 6
intellectual property protection 29, 35
interior designers 11
international competitiveness 55–6, 57
international trade 28–9
Isola, Maija 63

Jahnke, Marcus 181–2, 184

Kenricks 31–3
kilts 181–3
knowledge
  contextual 95–7
  tacit 119
knowledge assets 2
knowledge-based economy 55, 144
knowledge class 59
knowledge exchange 206
knowledge society 51

labour, division of 26–31, 117–18, 163
labour market 66
  local 131–3
Leathermarket 158
Legoretta, Ricardo 156
Leipzig 143
Lindfors, Stefan 57
Living Lab 71–2
LNG-fuelled ferries 212–13
local buyers 102–4
local buzz 96, 121
local-global divide, role of cultural intermediaries in bridging 106–12
local labour markets 131–3
locational patterns, of U.S. industrial design firms 81–91
logos 226–7, 228, 229, 236, 237
London 147, 150–8
Los Angeles 85, 145
low value-added goods 1

MAGIC 107–8
Manhattan 147
manufacturing
  arts and 37–8
  dynamics of global 2
Maritime Colours 214
maritime industry, Norwegian 195–217
  categories of design firms 208–9
  competences, organizational models, and market diversity 207–14

maritime industry,
  Norwegian – *continued*
  geography of   206–7
  product design examples   209–14
MaritimeMøbler   210
marketing   28, 175, 191
market segmentation   177
Marxism   59
mass media   31
mass production   28–31
McCloskey, Deidre   175
Melbourne   147
men's grooming products   180
meritocracy   66–7
metropolitan areas, *see* cities
Michigan   82
Milan   143, 147
Molotch, Harvey   145
Montreal   125
Montreal Collections   110–12, 113
Montreal fashion   93–114
  cultural intermediaries in   106–12
  independent stores and   102–6
  industry structure   100
  local advantage   102–6
  Montreal Collections   110–12, 113
  within North American marketplace   98–102

Napoleonic Wars   41
National College of Art and Design   36
national competitiveness, industrial design and   46–7
national consumer behaviour   3
national design identity   46–7
national economic policy, incorporation of industrial design into   35–40
national economy   23–4
natural landscapes   69
neoliberal discourse   52
Netherlands   224
networks   131–3
network society   51
neutral design   179
New economy   142, 144
new international division of labour (NIDL)   163

new media   152–3, 159–60
New York   82, 147
New Zealand   52
Nokia   54, 63–5, 69
North American Free Trade Agreement (NAFTA)   99
North American marketplace, Montreal fashion and   98–102
Norwegian Design Council   195, 207
Norwegian maritime industry   195–217
Novopen   237
Nuovo, Frank   64

occupations, rise of new   28–9
offshoring, of production   1–2, 117–18, 127–9, 133–4
Oregon   82
outsourcing, local   127–9

Paris   143
Partington, Angela   174
patents   11, 29, 30
Peel, Robert   38
Pennsylvania   82
performative capitalism   56
performative spaces   67–75
Petty, William   27–8
physical proximity   96–7, 121–2
Pippilotta   181
place   96–7, 121, 144–5
  in creativity discourse   67–73
planned obsolescence   93
Plup Water   57
policy environment   23–5
Porter, Michael   55
portfolio careers   65–6
post-corporate syndrome   159–60
post-Fordism   142, 144–5, 147–8, 149
postmodern architectural vision   70
poststructuralism   170
power tools   180–1, 184, 185
pre-industrial activity   144
price-based competition   1, 6, 29–30
private sector, policy environment and   23–5
product design   227–9, 236
product development   5

production
  consumption and 95–6
  divisions of 143
  factory 26
  Fordist 143
  mass 28–31
  offshoring of 1–2, 117–18, 127–9, 133–4
  removed from design 118–19
production-consumption divide 104–6, 113–14
production systems
  design-intensive 6
  geography of 199–201
product recognition 9
products, designer 3, 9–10
project-based design work, in Toronto 117–37
project teams 10
property market 148, 150
proximity 96–7, 133–4
Pugin, A.C. 37

quality 221–2, 231–6
Quebec City 125–6

recycling 13
Reform Act (1832) 38
regional competitiveness 55–6
regional development agencies 24–5
regional policy 13
relational proximity 98, 112–13
Renaissance 26
research and development (R&D) 10–11, 227–8
resource-based view (RBV) 6–7
responsibilization 66
resurgent city 141–4
retailers
  fashion industry and 94, 99, 106–12
  independent 102–6
return on investment (ROI) 230–1
Rhodes, Zandra 156, 158
Royal Dublin Society (RDS) 35–6
Royal Society for the Encouragement of Arts, Manufactures, and Commerce (RSA) 36–7, 41–2
RTI Sports 180

sales representatives 106–7, 110
San Francisco 85, 147
Sarpaneva, Timo 63, 69
Scandinavian Design 50
Scott, Allen 141, 146, 162
Scott, Linda 175
Seattle 85
Select Committee on Arts and Manufacture 39
self-care 66
self-employment 121
service economy 143
Shanghai 164
showrooms 108–9
Singapore 52, 164
Singer Sewing Machines 29–30
Slothfully 2006 185
small design firms 202–4
Smith, Adam 26, 28
social networks 132–3
social structure 60
Society for the Encouragement of Arts, Manufacturers, and Commerce (SEAMC) 36, 37
soft capitalism 57
Soho Works 33–5
space 73–5
spatial concentration, of U.S. industrial design firms 82–6
spatiality 143–4
spatial model of design 145–9
spatial proximity 96–7, 121–2
specialization 143, 204–5
Spivak, Gayatri 175
Stiletto 185–6
sunrise industries 144
sunset industries 144
supply-side perspectives, on U.S. industrial design firms 86–9
Swatch watch 8
Sweden 127

tacit knowledge 119
talent 68, 69–70
task specialization 143
technology 54, 68, 71–2
  creativity and 60–1
Temple of Industry 40
Third Italy 143

Tokyo 163
tolerance 67, 68, 72–3
Tompion, Thomas 27–8
Toronto 56
  project-based design work in 117–37
toys 178–9
trade shows 107–8
Turkey 1

ULSTEIN X-BOW 211–12
Umoe Mandal 214
UNESCO 52
unisex 178–9
United Kingdom, regional development agencies 24–5
United States
  Canadian exports to 99–102
  designers in 127
  industrial design firms in 81–91
universal design 11, 178
urban creativity paradigm 74

urban design 145
urban economy 119–22
user context 12

valued-added goods 1
Vancouver 125, 158–62
Venice 143
Victoria 126
Vik-Sandvik Industrial Design 210
Volkswagen 237
Volvo YCC 172, 186–91

watch industry 27–8
Watson, Tom 220
web design 227, 228, 230, 237
Wedgwood, Josiah 26–7
Windsor Castle 37–8
Wirkkala, Tapio 54, 63, 69
Worcestershire Exhibition 43
working clothes 179
world cities 55
World Trade Organization (WTO) 99